Turpentine Days, **Robert Butler 1996**
Plate 3 (Agricultural Heritage Series)
Sponsored by Florida FFA Alumni Association

Agriculture is important to our community, our state and the world. Agriculture is the number one industry in the state of Florida.

Reading is important in every walk of life. This book is donated to encourage students to read and develop respect and understanding for Agriculture.

The following Agriculture Industry groups provided this book for you on 4/18/06:

Sarasota County Cattlemen's Association
Sarasota County 4H Foundation
Sarasota County Fair Board Association
Sarasota County Farm Bureau
Sarasota Vo Ag FFA
Sarasota Vo Ag FFA Alumni

is proud to be the sponsor for

The Land in the Sunshine: A Look at Florida's Agriculture.

The proceeds from this book will benefit members of the Florida FFA Association.

A big thank you is in order to Mr. Archie Matthews from Alachua, Florida, for heading up this book project. For information about the alumni, he may be reached at 386-462-3016.

MISSION STATEMENT

We strive to be Florida's premier agricultural education advocate through FFA alumni support in our local communities. Our organization provides resources and support in the development of premier leadership, personal growth, and career success of FFA members.

VALUES

- We value the integral nature of FFA and agricultural education.
- We value agriculture as an essential part of society.
- We value diversity in serving all populations.
- We value development of the whole person.
- We value the impact of a teacher on a student's life.
- We value the community's support of agricultural education teachers.

THE
DONNING COMPANY
PUBLISHERS

THE LAND IN THE SUNSHINE

A LOOK AT FLORIDA'S AGRICULTURE

by

Archie Matthews

for

The Florida FFA Alumni Association

Greener Pastures, *Robert Butler 1996*
Plate 4 (Agricultural Heritage Series)
Sponsored by Florida FFA Alumni Association

To the overworked, underpaid, and too often unappreciated teachers of Vocational Agriculture who have inspired countless thousands of young people to assume roles of leadership, this book is affectionately dedicated.

The Donning Company Publishers
184 Business Park Drive, Suite 206
Virginia Beach, VA 23462

Steve Mull, General Manager
Barbara B. Buchanan, Office Manager
Richard A. Horwege, Senior Editor
Andrea L. W. Eisenberger, Graphic Designer
Stephanie Danko, Imaging Artist
Mary Ellen Wheeler, Proofreader
Lori D. Kennedy, Project Research Coordinator
Scott Rule, Director of Marketing
Travis Gallup, Marketing Coordinator

B. L. Walton Jr., Project Director

Library of Congress Cataloging-in-Publication Data

Matthews, Archie, 1954–

The land in the sunshine : a look at Florida's agriculture / by Archie Matthews for the Florida FFA Alumni Association.
p. cm.
Includes bibliographical references (p.).
ISBN 1-57864-244-2 (hard cover : alk. paper)

1. Agriculture—Florida. I. Florida FFA Alumni Association. II. Title.
S451.F6M28 2004
630'.9759—dc22

2003027241

Printed in the United States of America
by Walsworth Publishing Company

CONTENTS

FOREWORD

by DOYLE CONNER

Recent figures indicate that Florida has around forty-four thousand commercial farms, utilizing over ten million acres of land. These farms produce a staggering variety and quantity of products.

Florida's farmers lead the nation in production of citrus and sugarcane, are second nationally in the production of greenhouse and nursery products, strawberries, and tomatoes, and fourth in aquaculture and honey. Hundreds of thousands of Florida calves are shipped to feedlots to be turned into prime beef.

These are just a few of the facts about Florida agriculture. Florida's farmers and ranchers are some of the most productive in the world, providing safe and reliable food and fiber for consumers worldwide.

Florida farmers have always been innovators and experimenters. A closed door becomes a window of opportunity for them. The history of Florida agriculture is replete with crops and livestock that became unprofitable or unpopular to grow. Substitutes and improvements were found and will continue to be found.

The strength of Florida agriculture is in people. People from virtually every nation on earth have come to Florida to escape oppression, to seek opportunity, and to follow their dreams. Our state is richer for their experiences and spirit. It is my hope that Florida will continue to be a beacon to those who would come to plant deep roots in our soil and find fulfillment in their lives here.

The next generation of Florida agriculturalists will face challenges that their predecessors did not have to. The next generation will have to solve problems of environmental quality, water quality and quantity, the encroachment of urbanization, and the eternal problem of how to feed and clothe more people from less land. I am confident that the young people entering the field today will find the solutions to these problems just as those who came before them did. The challenge to bring about "better days through better ways" is still there.

As a former FFA member, a past State and National FFA president, Florida's Commissioner of Agriculture from 1961 to 1991, and as one who has a deep, abiding, and lifelong love of agriculture, it gives me great pleasure to contribute to this book.

PREFACE

FLORIDA AGRICULTURE IS ROOTED IN RICH HISTORY

by CHARLES H. BRONSON,
Commissioner of Agriculture

When looking at the accomplishments of Florida agriculture today, it is fitting to look also at the long history of our state and the traditions of farming that have brought it to international prominence.

As a fifth-generation Floridian who has worked closely with the Future Farmers of America, I am honored to have been asked by the Florida FFA Alumni Association to contribute to a book that so impressively portrays our state's farming history. I am also very proud that my family has played a part in that history. My ancestors came to the New World in 1635, less than a generation after the Pilgrims landed at Plymouth Rock, and they actually worked with the first cattle brought to Florida through South Carolina. The town of Bronson in Levy County is named after my great-great-grandfather, who settled in Florida in the 1830s. My family have been ranchers in Florida ever since.

Florida has come a long way in its agricultural development and rightly deserves its status as one of the leading agricultural states in the country. Nearly one-third of its 34 million acres are

Shang Bronson, father of Florida Agriculture Commissioner Charles H. Bronson, herds cattle in Kissimmee in the 1960s.

PHOTO COURTESY OF CHARLES H. BRONSON

in use by 44,000 farmers who employ more than 94,000 farmworkers to produce over 280 crops that generated $6.42 billion in farm receipts in 2001 with an annual overall economic impact estimated at more than $50 billion.

Famous for citrus production, Florida's growers cultivate more than 103 million trees on over 797,000 acres, providing more than 78 percent of the country's citrus.

Florida's greenhouse/nursery sales of more than $1.6 billion make it the second leading horticulture state in the nation. In 2002, floriculture sales reached $860 million, and foliage plant sales topped $459 million.

Farmers produced almost $2.9 billion of commercial vegetables, fruits and nuts in Florida during the 2002 season. Florida ranks second in the country in the production of fresh fruits and vegetables and provides approximately 80 percent of those grown in the United States during January, February, and March. Florida also leads the nation in sales of fresh tomatoes, sweet corn, snap beans, cucumbers, radishes, watermelon, and eggplant. In 2002, agricultural exports reached $1.267 billion, ranking Florida fourteenth in the United States.

Florida's livestock industry grossed $1.239 billion in 2002 with cattle and calf sales of more than $333 million, dairy sales of $356 million, and chicken and egg receipts of $376 million.

The state averages an annual seafood harvest of more than 96 million pounds that has a dockside value greater than $189 million, which consistently ranks it among the top dozen states in the United States. In 2001, tropical fish were the number one commodity in Florida's aquaculture industry, with sales of $42.4 million out of a total of $99.5 million of aquaculture products sold.

Florida's 16.5 million forested acres make up almost half its total acreage, and in 2000, the state's 14.9 billion cubic feet of standing timber produced $429 million of harvested timber and manufactured products with a value of $8.6 billion.

Florida Agriculture Commissioner Charles H. Bronson herds cattle at the C. H. Bronson Ranch in Osceola County in 1994.

PHOTO COURTESY OF CHARLES H. BRONSON

There are about 300,000 horses in Florida, and almost a quarter of a million Floridians are involved in the state's $6 billion horse industry, which has already produced such great champions as "Silver Charm," the 1997 Kentucky Derby and Preakness winner, and "Skip Away," who was the 1998 North American Horse of the Year.

Throughout the state's history, Florida farmers have shown themselves to be outstanding providers for a growing population while being environmentally aware of the limits of the natural resources on which they depend. They have also proven themselves to be competitive both nationally and internationally and open to technological innovation that will keep them among the best producers in the world. Work by groups like the Future Farmers of America will ensure that Florida's agricultural future will be as rich as its past, and works like this book are a reminder of just how rich that past has been.

ACKNOWLEDGMENTS

The author recognizes that this book would not have been possible without the invaluable assistance of many people, including the officers and directors of the Florida FFA Alumni Association, the officers and staff of the Florida FFA, the Florida FFA Foundation, and the many FFA members and FFA advisors who have provided photographs and information. Past and present staff members of the Florida Department of Agriculture and Consumer Services have been invaluable for their advice and support. I would also like to thank my wife, Emelie, for her patience and support during this project, which consumed more of us and our time than we imagined it would. Thanks to Bernie Walton and Richard Horwege of Donning Company Publishers for your advice, intelligent answers to my stupid questions, and most gratefully for your patience through what must have seemed like endless delays.

To those who have not been mentioned, the oversight is mine alone, and your contributions are nonetheless appreciated.

Azaleas in bloom.
PHOTO BY THE AUTHOR

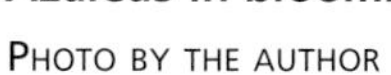

INTRODUCTION

A BRIEF HISTORY AND OVERVIEW OF FLORIDA AGRICULTURE

The first European to set eyes on what is now Florida named it *La Florida*, "Feast of Flowers." Indeed, with the year-round mild temperatures and abundance of rainfall, there is hardly a day when something isn't blooming. These same mild temperatures allow for a twelve-month growing season, with the planting and harvesting of crops moving up and down the state with the seasons.

The Native Americans inhabiting Florida practiced subsistence agriculture long before the first European settlers landed on these shores. Some estimates place Native Americans in Florida over twelve thousand years before the first Europeans came. While primarily hunter-gatherers, they had also learned from experience how to produce food from the land, growing crops of corn, beans, and squash. When the first European settlers came, bringing livestock and practices that the Native Americans had never seen, they were quick to adapt to these new aspects of agriculture.

Jackfruit, a tropical exotic fruit, is now grown in South Florida.

Photo courtesy of Wayne Worthley

Livestock that strayed or escaped from their European masters were seized by the Native Americans. Many of them became accomplished stockmen, tending large herds.

Some of the first Spanish settlers who moved to Florida to farm hoped they could remake Florida into New Spain, with groves of olives and vineyards of wine grapes such as they knew already. These hopes were quickly dashed when it was found that European grapes were not resistant to native diseases and pests and that Florida was much too humid for the olive to flourish.

Attempts were made to open the interior of Florida for settlement through the granting of large tracts of land by the King of Spain to those who would come and spread civilization within Florida. These boundary lines still appear on modern maps. The spread of the Spanish Missions through the interior of Florida had a great influence. The priests at these missions sought to convert the Native Americans to Christianity and to teach them practical skills.

Spain ceded Florida to the United States in 1819, and it was established as a territory the following year. The territorial period was marked by hostilities with the Seminole Indians, who were Creeks from Georgia and the Carolinas who had moved in during the period of European settlement. The native Timucuans, Apalachees, and Caloosas had been devastated by diseases to which they had no resistance, contracted from the European settlers. Continual skirmishing and conflict reduced their numbers even more. It is estimated that 90 percent of the native population lost their lives as a direct result of the European settlement of Florida. The Seminoles moved further into Florida, evading the troops sent to remove them. They finally settled in the vast swamplands of the Everglades around Lake Okeechobee. The efforts of armed troops were unable to round up and remove many of the Seminoles who had a firm hold in this treacherous land. The Seminoles are the only Native Americans to have never

Spanish explorers brought cattle like these on their early voyages.
PHOTO COURTESY OF GARY LEE

Macadamia nuts, once exclusively imported, are now grown in Florida.
PHOTO COURTESY OF WAYNE WORTHLEY

signed a treaty with the United States, and they remain on their adopted land, growing crops, producing fine cattle, and are a vital and respected element of Florida's diverse population.

Florida's unique climate is largely responsible for its vast diversity of agriculture. The northern Panhandle and Peninsula are the home to traditional agriculture as is practiced by neighboring states. As one moves down the Peninsula, the climate becomes more and more tropical, and many crops flourish there that cannot grow any other place in the United States.

Florida agriculture is a constantly changing mosaic. Crops that once formed the backbone of farm communities are gone, never to return. The groves of tung-oil trees that once grew in North Florida are gone, their natural industrial oil replaced with synthetics. The tobacco grown under the shade of cheesecloth to make cigar wrappers that once defined the economy of a region is gone, a victim of the prohibitive cost of hand labor. Indigo, grown for the rich blue dye it produced, is long gone, replaced by chemical dyes. Hemp is gone, its fiber replaced with synthetics and with cheaper foreign imports. No longer are groves of native and planted pine timber slashed to collect resin to distill into naval stores. Some of these products have been replaced with synthetics, and resin is recovered from stumpage and the waste of paper mills.

Livestock production as well has seen its share of changes. The mule was once bred and used extensively in Florida, but now exists only as a novelty. Florida native sheep had their heyday, but have never even been heard of by most Floridians. The "cracker pony" still exists, but is far outnumbered by sleek thoroughbreds and quarter horses, and every other breed of horse. The "cracker cow" that formed the foundation of Florida's vast cattle industry, remains only as a dim shadow of what it once was, supplanted by other breeds.

Versatility: This young farmer is using a bulk tobacco curing barn to dry sweet potatoes.
PHOTO COURTESY OF SHIRLEY CARTE

Longleaf pine seedling. Once the heart and soul of the timber and turpentine industries, the longleaf pine is being restored to some of the land it once dominated.
PHOTO BY THE AUTHOR

In spite of these losses, there has been no net loss of diversity in Florida agriculture. Each crop or method that is outmoded is replaced with something else. There are currently in excess of 250 commodity groups, each representing a specific product, and there are some commodities so new that there has not been time to form an organization to represent them. New crops and specialty livestock are introduced each year, some to cater to the tastes of diverse ethnic backgrounds, and some because they are unique, new, and exciting.

The scope of Florida agriculture is limitless. There has never been a more exciting time to be involved in this most important profession. New methods, technology, biological advances, and things only dreamed of by the writers of science fiction await the farmer and rancher of the next generation. The "gee whiz" technology that is today's cutting edge will become tomorrow's dinosaur, relegated to the dustbin of history, surpassed by even greater achievements. The task of feeding and clothing the ever-growing population of the world will never be easy. Future generations will be forced to produce more food and fiber on less land, and more marginal land at that. The public must be educated to the reality that producing food, fiber, paper, lumber, and countless other necessities of life is an unrelenting task, and that the food they find readily and cheaply on the shelves of their favorite grocery store gets there only after a lot of hard work, time, and expense, and is not produced in the back room. The consumer must understand the total effect that agriculture has on the economy, that each job in production agriculture generates at least fifty jobs down the line in processing, distribution, sales, and other aspects of wholesale and retail

Fruits and Vegetables grown at the Live Oak Research and Education Center.
PHOTO BY THOMAS WRIGHT

business. They must also be shown that the farmer and rancher produce and market products while assuming all of the risk. There is almost no "safety net" for natural disasters, recession or depression in the national and world economy, no insurance that can adequately compensate for the loss that every producer knows he will have to incur someday. The producer takes the smallest share of the food and fiber dollar. Every step along the way from the farm to the consumer adds value to the product, but almost none of this goes back to the producer.

Florida farmers and ranchers will continue to overcome these challenges. Agriculture is more than a profession, it is a way of life. Most people involved in agriculture would rather do that than anything else. The sense that what they are doing is important and essential is a driving factor. Whether the individual is a producer, teacher, researcher, salesman, processor, or volunteer leader or coach with a local FFA chapter or 4-H club, everyone involved knows that what he or she does is important to the life of our nation and world.

Innovation: An orange and blue hibiscus, developed by researchers at the University of Florida.

PHOTO COURTESY OF BRENDA SULLIVAN

The rolling hills of North Florida.
PHOTO BY THE AUTHOR

CHAPTER 1

FARMS AND LAND

According to the latest available figures, Florida has about forty-four thousand commercial farms, totaling 10,200,000 acres. Florida ranks twenty-second in the number of farms in the United States and thirtieth in land in farms.

These bare statistics do not begin to tell the story of Florida's farms and ranches. If you were asked to describe a typical Florida farm, you would find that this is an unanswerable question. There is so much difference between a farming operation in the Panhandle and one in the Lower Peninsula that you would be unlikely to think they existed in the same country, much less the same state. The rolling hills of North Florida, with their red clay soils, give way to sandy loams as you move south, and then turn into muck soils, pure organic matter, incredibly fertile, but fragile. These huge regional differences have created agriculturalists

Much of Florida land is characterized by an abundance of water as in this scene from a Florida wetland.

PHOTO BY TONY LAYNE COURTESY OF DESERET CATTLE AND CITRUS

who have adapted to the characteristics of the soil and climate of their farms.

The heavier soils of the Panhandle counties are reminiscent of the soils of neighboring Alabama and Georgia, and a wide variety of traditional crops and livestock are produced on them. Corn, peanuts, soybeans, rye, oats, cotton, tobacco, and all of the grasses for grazing and hay production are grown there, and grown well. The soils have a high-clay content, and must be treated accordingly. Where the land slopes to a significant degree, the fields must be terraced to slow down the runoff from rainwater or irrigation that would carry away the precious and irreplaceable topsoil. This terracing allows the water to slow down and soak into the soil rather than just run over it.

The lighter, sandier soils of the Upper Peninsula are productive soils also, producing all of the crops listed for the Panhandle, with perhaps the exception of cotton. Cotton was once king far down the Peninsula, with areas such as Alachua County producing the highly prized Sea Island cotton. The devastation of the boll weevil ended the reign of King Cotton for most of the state. Chemical controls and a highly successful eradication program have caused cotton acreage to increase, but only the Panhandle counties are within reach of the ginning facilities necessary to handle the crop. Attempts have been made within recent years to restore cotton to areas where it was once grown, and while production was good, the cost to haul the raw cotton to the nearest gins made the venture impractical.

Within the Peninsula are areas of what are known as flatwoods. These areas are characterized by land that is very level and has a high water table, acidic soil, and high organic content. These soils are very productive, and while virtually all crops grow well here,

Much of Florida's farmland is rapidly being converted to housing as cities continue to grow and residents seek the quality of rural life.

PHOTO BY THE AUTHOR

As this photo shows, much of Florida land is flat as a tabletop, nearly featureless except for what has been planted.

PHOTO COURTESY OF HOLT FARMS

much of this acreage is used for vegetable production. Squash, cucumbers, peppers, potatoes, eggplant, and green beans are some of the primary crops produced here.

The Central Florida Sand Ridge does not sound like a very productive soil to the uninitiated. One is used to associating sand with beaches, as a part of concrete, and at playgrounds. Floridians are proud of their sand and many of Florida's soil types have sand as part of their name. The state soil is Myakka Fine Sand, and is

Cotton still thrives in the Panhandle counties of Florida.

PHOTO COURTESY OF RALPH YODER

Seed pieces of sugarcane are planted by hand in Everglades soil.

PHOTO COURTESY OF HOLT FARMS

unique to Florida. Sandy soil can be well or poorly drained, depending on its location. Sandy soil can be highly productive, or a poor producer. The sands of the Ridge, as it is known, are some of the finest soils for producing citrus in the world.

Farther south, the countless lakes in the central region of the state drain into an area known as the Everglades. The Everglades is a vast slow-moving river, eventually emptying into Florida Bay. Drainage projects in the last century opened up much of this land to farming and ranching. These muck soils are mostly organic matter, built up over millennium beneath the waters of the Everglades. Muck soils are unbelievably fertile.

Much of this region is far enough south to avoid most of the devastating freezes that make it to Florida, so there is something growing there year-round. Huge tracts are devoted to sugarcane. Rice, sweet corn, and all types of vegetables are grown here. Muck soils are found throughout Florida, and are highly productive lands. Conservation efforts have reflooded some of these areas, and the wetlands originally found there are being restored. Wetlands provide natural filtration for watersheds, and the importance of these areas has been recognized.

These generalized areas of soil do not begin to tell the picture of the diversity of soil types in Florida. It is common for very disparate types of soil to lie very near to each other. A soil survey of even a small farm will generally show numerous types of soil, each with its own unique characteristics. The USDA through the Natural Resources Conservation Service, formerly the Soil Conservation Service, has mapped the soils throughout Florida, and maps and specifications of these various soils are available from NRCS offices. Local Soil and Water Conservation Districts in partnership with Water Management Districts seek to educate the public to the absolute necessity of preserving and protecting the precious land. Once topsoil is lost to erosion, it is gone forever. It is the most valuable resource on the planet. Our land will only support us as long as we provide the stewardship to maintain and preserve it.

Even the lakes, rivers, and oceans become the farms of the aquaculturist.

PHOTO COURTESY OF KATHRYN MCINNIS

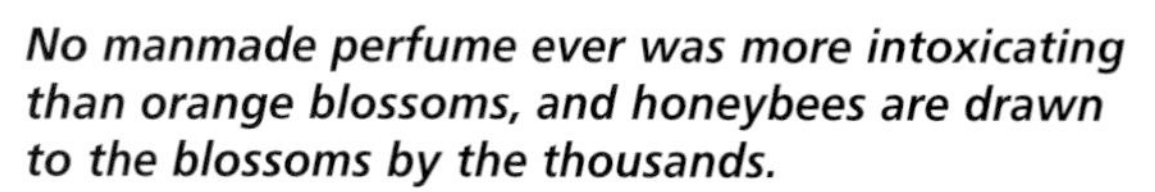

No manmade perfume ever was more intoxicating than orange blossoms, and honeybees are drawn to the blossoms by the thousands.
PHOTO BY THE AUTHOR

CHAPTER 2

CITRUS

Florida leads the nation in citrus production. Three-quarters of the citrus produced in the United States is produced in Florida. It is the defining commodity produced in the state. When one thinks of Florida agriculture, the connection with citrus production is inescapable. Nearly eight hundred thousand acres are planted to citrus. Thirty-two of Florida's sixty-seven counties are identified as citrus production areas.

The early Spanish explorers and settlers brought the first citrus trees to Florida, and like many other plants it found a congenial home. The settlers took citrus trees with them as they moved into the state, and the Native Americans planted them near their settlements. Seeds were spread naturally by the actions of birds and other animals, and citrus trees were eventually found growing wild in virtually all parts of the state.

Researchers and growers introduced and developed new varieties of citrus and better

This many blooms on mature orange trees foretell a bumper crop.
Photo by the author

methods of production. The seedling tree gave way to grafted stock, producing sooner, with more disease resistance, greater production, and other desirable traits. Plant breeders crossbred and hybridized new varieties, producing fruit that was superior in size, color, flavor, and productivity. Many varieties of citrus that are common today had their beginnings in Florida groves and research centers. Common varieties of Florida citrus include Navel, Valencia, and Hamlin oranges; Ruby Red, Thompson, and Duncan grapefruit; and tangerine types such as Tangelos and Temple oranges. Lemons and limes are grown in much smaller numbers, around a quarter million boxes each, annually.

The citrus production area was initially in the northern Peninsula, but only a shadow of that production remains. Some small growers produce limited amounts of early, cold hardy citrus, such as the Satsuma and the incomparable Chinese Honey tangerine. A series of devastating freezes, which began in the 1890s and continued sporadically and unpredictably throughout the twentieth century, moved the planted acreage south, in an attempt to find an area without freezes, but with suitable soil types. Areas in North Florida, such as Mandarin and Orange Heights, whose names reflect their former dependence on citrus, are now many miles from any commercial citrus plantings. Even the once burgeoning citrus production area in Central Florida has had its producing acreage greatly reduced by freezes, with many growers unable to replant because of the cost of clearing the dead trees and the increased value of the land for development. This area has some of the finest soil for producing citrus on the planet, but Florida's ever-increasing population has converted much of this land into

The proof is in the picking. This fruit is ready for the processing plant.

PHOTO COURTESY OF CRITTENDEN FRUIT COMPANY

Late season grapefruit ready for harvest.
PHOTO BY THE AUTHOR

commercial and residential development. Many people now live and work on land that once produced world-famous citrus fruit. Much of Florida's citrus is directed to processing. Production of juice takes around 90 percent of Florida oranges. Millions of gallons of this juice are concentrated and frozen. Every type of citrus, except lemons, is turned into concentrate. Florida researchers developed and perfected the concentration process as a way to recover fruit that had been damaged by freezing and was worthless on the fresh fruit market. Millions of gallons of citrus concentrate are produced for nationwide distribution, and exports exceed 17 million gallons of frozen concentrate annually.

Much like the pig, about which meatpackers used to say that you can sell everything but the squeal, the byproducts of citrus juice production have considerable value. After the juice is squeezed out, the remaining pulp, peel, and seeds are processed into

Young citrus trees coming into good production.

PHOTO COURTESY OF CRITTENDEN FRUIT COMPANY

Trailers of fruit await their turn at an orange juice concentrate processing plant.
PHOTO COURTESY OF GARY LEE

nutritious, low-cost cattle feed. The oil extracted from the peel is used in the manufacture of paints and varnishes. Citrus molasses can be fermented into alcohol, or used in cattle feed. Citrus byproducts show up in a variety of foods, cosmetics, insecticides, and other consumer products. Florida beekeepers sell world-famous orange blossom honey. If the sweet perfume in the air of a blossoming orange grove could be captured, truly nothing could be said to go to waste.

The fields are so vast that United States Sugar has its own internal rail system to deliver cane to the sugar mills.

PHOTO COURTESY OF UNITED STATES SUGAR CORPORATION

CHAPTER 3 SUGARCANE

Florida is one of only four states that produces sugarcane commercially, and is the largest producer of sugarcane and cane sugar. Florida produces over 16 million tons of cane annually, more than 25 percent of the nation's sugar supply.

Sugarcane, like many of Florida's other crops, had its beginnings in the earliest European settlements. Spanish explorers brought sugarcane to the West Indies and then to Florida. Early efforts to produce sugar were semisuccessful, but were hampered by inadequate knowledge of sugarcane's requirements of soil and climate. Attempts to produce sugar were made at Cape Canaveral. *Canaveral* is a Spanish term meaning "cane field." Other attempts were made at Homosassa, Port Orange, and New Smyrna. What remains of the David Levy Yulee sugar mill has become a state historical site, and provides a look at what one of these early sugar mills looked like.

A "one-horsepower" cane mill, typical of what was once commonly found in every farm community.
PHOTO BY THE AUTHOR

Small-scale production of sugar and syrup was conducted by individual farmers, often producing only enough for the needs of the family, with perhaps a little left over for sale or barter. This small-scale production is still carried on by some farm families, not for survival as in earlier times, but to retain this link to the past.

This closer view shows cane being crushed or "ground" between the cast-iron rollers of the mill.
PHOTO COURTESY OF JACK WILLIAMS

Both man and mule have to work pretty hard operating a cane mill.

Photo courtesy of Jack Williams

Families work together to grow, harvest, grind the cane, and produce sugar or syrup, more often syrup. The "cane-grinding" is still a social event in a number of communities, with several generations working together. Historical sites such as the Dudley Farm Historic State Park near Gainesville and the Morningside

Jack Williams, retired FFA advisor and an accomplished master syrupmaker, tests the syrup to see if it is done.

Photo courtesy of Jack Williams

The cane leaves, called "fodder," are burned off prior to harvest. Here a worker is lighting a field.
PHOTO COURTESY OF UNITED STATES SUGAR CORPORATION

Nature Center in Gainesville conduct annual cane-grindings and syrup cookings to illustrate this aspect of pioneer life.

Large-scale sugar production took off with the vast plantings of cane in the muck soils around Lake Okeechobee. Muck soil is mostly organic material, and is extremely fertile. Nearly half a

Mechanical harvesters will move into the cane field after the fire.
PHOTO COURTESY OF UNITED STATES SUGAR CORPORATION

A mechanical cane harvester at work.

PHOTO COURTESY OF HOLT FARMS

million acres are devoted to this crop, supporting six sugar mills and two refineries.

Few people realize that sugarcane is merely a giant grass. Most everyone can look at it and see in its growth habits and appearance a close kinship with bamboo and the cornstalk. Its value is in its ability to convert soil nutrients, water, and sunlight into sugar. Plant breeders have successfully crossbred and hybridized sugarcane to increase this trait, while improving disease resistance, hardiness, and milling characteristics. Sugarcane is not grown commercially from seed. Sections of the stalk are planted, and a new plant arises from the "eye" located at the intersection of each of the joints. The cane stalks are harvested each year.

Formerly, the crop was harvested mostly by hand, but improvements in mechanical harvesting have entirely replaced the backbreaking, dangerous work of cutting cane by hand. Cane fields are burned prior to harvest to remove excess foliage, and the cane mills run night and day until the crop is harvested, milled, and reduced to raw sugar. The refining of the raw sugar takes place at a more sedate pace. Raw sugar is stored in warehouses, and looks like large piles of sand, and is handled with some of the same type of heavy equipment used to move and handle sand. It is removed and refined as need occurs. The sugarcane industry is a major factor in Florida's agricultural economy, bringing in nearly $450 million in sales, providing in excess of forty thousand jobs, with a total impact on the state economy of nearly $2 billion.

Cane fields stretch to the horizon in the Everglades.
PHOTO COURTESY OF UNITED STATES SUGAR CORPORATION

This mature field of sugarcane is in bloom and will produce seed. The seed is tiny, a spoonful contains a couple of thousand. Cane is planted by burying sections of the stalks in rows.
PHOTO COURTESY OF HOLT FARMS

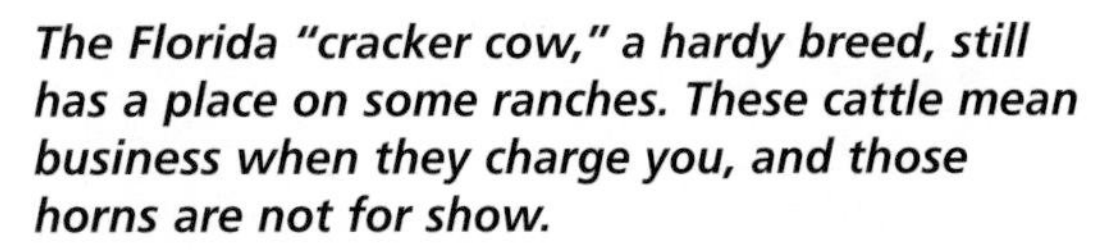

The Florida "cracker cow," a hardy breed, still has a place on some ranches. These cattle mean business when they charge you, and those horns are not for show.

PHOTO BY THE AUTHOR

CHAPTER 4

CATTLE

The cattle that Ponce de Leon brought with him on his first voyage in 1513 were not the fine, large, well-muscled or large-uddered cattle that we see today. These Spanish cattle were much smaller, producing little milk, and nothing resembling the tender, marbled beef that we enjoy today. Many of them wandered away and became wild, living off the land in the thick Florida scrublands. These early cattle fell prey to many tropical pests and diseases that were unknown to them in their native country. Those that survived the rigors of this cruel and merciless process of natural selection and survival of the fittest, developed strong instincts for survival, and the ability to thrive on the poorest of pasture. They also usually calve very easily, and are nearly disease free, valuable traits that the industry hopes to gain from these cattle. These "cracker cattle," as they have become known, are small, aggressive, very intelligent, and very protective of their calves and the entire herd. They share

A "cow whip" is an essential tool of the Florida cow hunter.
PHOTO COURTESY OF LIGHTSEY CATTLE COMPANY

common roots with the famous Texas longhorn cattle, but are recognized as a separate breed, and have their own registry. A number of Florida ranchers maintain small herds of these wiry little cattle. Lake Kissimmee State Park near Lake Wales has a herd of cracker cattle, and has re-created an early Florida "cow camp." The

When driving home a herd, the cow whip is used to direct the cattle, not to beat them. Notice the cow dogs following behind the horses. They are ready to turn back cattle that try to escape into the scrub.
PHOTO COURTESY OF LIGHTSEY CATTLE COMPANY

Start of another day working cattle.

PHOTO COURTESY OF LIGHTSEY CATTLE COMPANY

Dudley Farm Historic State Pake near Gainesville also has some cracker cattle.

The American cowboy began in Florida, not in the Western states as Hollywood would lead us to believe. Florida cowboys were known as "cow hunters" because much of their time was spent

A cow hunter on a cracker horse heads out to the range followed by a trained cow dog.

PHOTO COURTESY OF TRUDY TRASK

Driving cattle through one of the open prairies along the St. Johns River.

PHOTO COURTESY OF TRUDY TRASK

Cattle are branded for identification.

PHOTO COURTESY OF LIGHTSEY CATTLE COMPANY

searching for rather than herding cattle. The cow hunter's essential tools were his horse, his "cow whip," and his dog. The cow hunter's dog was not some purebred, well-trained, registered canine with an impressive pedigree, but was of indeterminate, and often undeterminable breed, with the instincts to find cattle, and avoid such perils as panthers, rattlesnakes, and alligators. The cow hunter's dog was sent into thickets, cypress domes, and palmetto heads, which were impenetrable to a man on horseback or on foot, to chase the cattle out so they could be branded or marked, doctored, and sold.

Florida's beef industry was vital during the War Between the States. Florida cattle were driven overland from the central prairies to railheads in North Florida and South Georgia. Florida was the major supplier of beef to the Confederacy. Many Confederate soldiers on the battlefields of Virginia and Tennessee were fed on Florida beef.

A Brahman with an oddball set of horns gets vaccinated.
PHOTO COURTESY OF LIGHTSEY CATTLE COMPANY

This Florida cow hunter is in his element. These cattlemen are as tough a breed as the cattle they work and the horses they ride.

Photo courtesy of Lightsey Cattle Company

The primary market for Florida cattle before and after the war was Cuba and other Spanish possessions in the Caribbean. Cattle were driven to such Gulf ports as Tampa and Punta Rassa, loaded on ships, and sent to these lucrative markets. Ranchers and dealers liked dealing in the Spanish cattle trade because many of the buyers paid

Herefords are one of the British cattle breeds that has been very successful in Florida.

Photo courtesy of Andrea Stevenson

Scrub cattle in Florida scrubland. The influence of several breeds can be seen in these cattle.

PHOTO COURTESY OF LIGHTSEY CATTLE COMPANY

in gold, and some of these ranchers and dealers had gotten stuck at the end of the war with worthless Confederate notes.

The cattle industry continued to improve. Mandatory dipping programs, mosquito control, and drainage projects did much to eliminate insect-borne diseases. Scientific breeding, research, and appreciation for Florida's unique climate have all contributed to the success of the cattle industry. Breeds from other parts of the world have been introduced here, both to increase production and to breed in resistance to the tropical heat. The introduction of Brahma cattle has been significant. This breed originated in India and has characteristics that help with resistance to heat. Crossbreeding with the more temperate breeds such as Angus and Hereford has improved the ability of Florida's beef cattle to cope with higher temperatures and humidity. Breeds from Africa have been introduced to diversify the genetics with the goal of producing cattle better adapted to Florida's climate.

Foreign breeds, such as this Watusi steer from Africa, are brought in and tested to see if their genetics offer any advantages to Florida cattle.

PHOTO COURTESY OF LIGHTSEY CATTLE COMPANY

Where once all cattle foraged on unfenced, unimproved, open range, most cattle are now kept on improved pasture. The introduction of Bahia, Coastal Bermuda, and other grasses dramatically improved the productivity of Florida pastures and hayfields. More cattle per acre are produced in Florida than in any other major cattle-producing state. The mild winters allow cattle to be pastured nearly year-round, unlike some of the Western and Northern states that may be dependent on hay and other supplemental feeds for half or more of the year. Florida now is one of the top five states in cattle production. Many Florida cattle are shipped to the Western states to be fed out in feedlots where more grain is available.

Florida cattlemen have always been at the forefront of breeding and development. Breeds have been introduced from every part of the world, and Florida breeders have taken desirable traits from many of them. International buyers come to Florida to purchase superior bloodstock to take back to their home countries.

Brangus, a cross between the Angus and the Brahman cattle, is now recognized as a distinct breed.

PHOTO COURTESY OF TOM AND SHIRLEY CARTE

In the modern milking parlor, the dairyman works in a pit with the cows on an elevated platform on either side.

PHOTO COURTESY OF D & D DAIRY

CHAPTER 5 DAIRY

Milk, butter, cheese, yogurt, and don't forget ice cream would not exist were it not for dairies. The production of milk is as old as agriculture. Early herdsmen would milk their livestock to provide this nutritious food. The Old Testament refers to the Promised Land as a land of milk and honey, testimony to the recognized value of these foods. The dairy cattle of today evolved through direct breeding from the general-purpose cattle of long ago. It was recognized that some cattle produced more milk than others, and these were selectively bred with other superior individuals, and the dairy breeds we recognize today came about. Like all aspects of agriculture, the dairy industry has undergone tremendous change. Dairying has been consolidated from the hundreds of small producers that once supplied the market to smaller numbers of large-scale operations. Once every farm kept a few milk cows to meet the needs of the family, but this is a rare practice anymore.

Some dairymen use Jersey bulls with Holstein heifers to keep birth weights low for their first calves.

Photo courtesy of Sonny Register

The breeds that once dominated the industry, such as Jersey, Guernsey, and Brown Swiss, have been largely supplanted by Holstein. The Holstein produces larger quantities of milk and thus generates more income for the dairyman.

The industry too has made great advances. Where it once took large numbers of workers to milk the cattle, now automatic milking machines are used. It was once common practice to wash the cow prior to milking, but research has shown that if only the teats are thoroughly cleaned and nothing else, there is less chance

Dairymen used flat barns for decades. The cows were washed with a large hose, and the dairyman had to perform back-breaking labor to attach the milking machines.

Photo courtesy of Douglas Register

of bacteria and contaminants entering the milk. The cows have computer chips attached to them, and scanners read these chips as the cow is milked, giving the dairyman a record of production and animal health. At the end of each milking, the dairyman can download all of this information and know which cows need some attention to their health, or when their production has begun to lag. Once the cows are milked, the milk is piped directly into refrigerated storage tanks where it is held until the milk tanker arrives, transferring it to a refrigerated truck to be sped to a processing plant to be prepared for the consumer. Once the milking shift is completed, the milking machines and their interconnecting plumbing are flushed with cleaning solutions and rinsed out, ready for the next shift.

These cows have been washed by an automatic computer-timed washer, and are waiting to enter the modern milking barn.

PHOTO COURTESY OF D & D DAIRY

Twelve cows at a time can be milked in this up-to-date, computer-assisted milking parlor.
Photo courtesy of Douglas Register

Milk and milk products are found in a variety of forms in the grocery store. Whole milk, reduced-fat milk, no-fat milk, lactose-free milk, chocolate milk, half-and-half, heavy cream, whipping cream, and buttermilk are all examples of fluid milk products on the shelf, in sizes ranging from half-pints to gallons. Butter, cheese in its endless variety, yogurt, cottage cheese, and ice cream are all familiar forms. Milk is also condensed or evaporated and canned for long-term storage, as well as other packaging that requires no refrigeration.

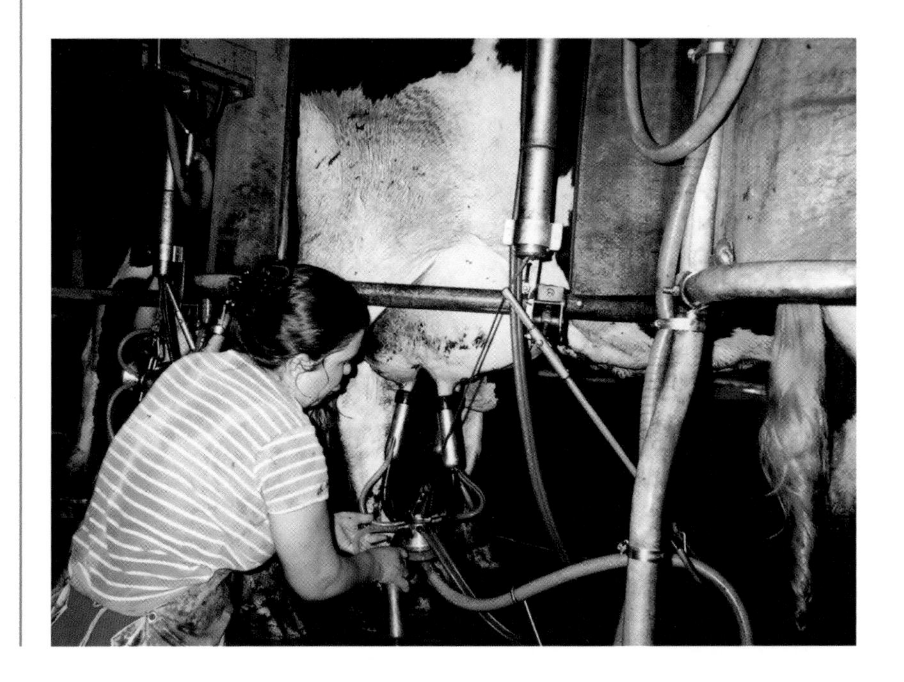

The dairy technician's task of attaching the milking machine is much easier now. She only has to lean over to attach the milking machine, and the machine detaches itself when milking is finished.
Photo courtesy of D & D Dairy

Cows can be held up after milking for medication if needed, or for artificial breeding if it is time.

PHOTO COURTESY OF DARRYL REGISTER

Florida dairymen have found ways to cope with the less than ideal climate for milk production. Research has found the best breeds for milk production in a hot climate. It has also found the right nutritional formulas to boost milk production. Florida dairy barns and sheds have large fans and evaporative coolers running continuously through the warmer months to help keep the cattle comfortable. Shade, whether natural or manufactured, is an important factor in dairy management.

The challenges facing the dairy industry include noise, odor, and fly control as more and more nonfarm people move into the country. There is the additional problem of waste management. The concentration of large numbers of livestock into relatively small areas can cause unwanted nutrients to migrate into groundwater

Cows rest in the pasture between milkings.

PHOTO BY THE AUTHOR

if not properly managed. Manure lagoons with impermeable liners, prevention of runoff, channeling the wastes back into crop fields to be taken up by growing plants are all ways that dairymen utilizing best management practices are attempting to address this problem.

Many things have changed in the dairy industry, but one thing never will. Those cows have to be milked twice a day, every day, in fair weather or foul. The dairyman makes a commitment that hardly anyone else can or will.

Milk is chilled and held in these six-thousand-gallon tanks until the bulk tanker truck arrives. Calculate how many cookies you would need if you drank that much milk.

PHOTO COURTESY OF DOUGLAS REGISTER

CHAPTER 6 HOGS

Like other livestock, hogs came with the early settlers, and like other livestock, some of them escaped and established populations in the wild. These hogs mixed with other feral hogs as they escaped, and a somewhat mongrel breed arose. The "piney woods rooter" was not very large, had stiff, coarse hair, the males grew elongated lower incisors known as "tushes," and was a formidable foe in the Florida woods. The ferocity of the wild hog cannot be overestimated. They formed into herds, and could take on any other animal in the woods. They could fall prey to well-armed men, but those men had to be skilled or they could find the tables turned on them very quickly. Later generations of sportsmen released the wild Black Russian hogs into the Florida woodlands to interbreed with the feral population, a move that increased their survival instincts, but did not improve their temperament. Feral hogs lived off of the land. The native oaks provided acorns, there were roots

The science of artificial insemination allows the swine producer to breed to the best boars in the nation.

PHOTO COURTESY OF TORI LYONS

to be grubbed up, and other wild forage, as well as the fields of the Native Americans and the early settlers.

Florida has many wild hogs today, in such large numbers that they become a nuisance, and there are active programs to hunt, trap, and remove them from sensitive lands. The early settlers who raised hogs used the bounty of the land to fatten them. The acorns, roots, and other forage were collectively known as "mast." Hogs were marked as to ownership at an early age, usually by cropping, notching, or splitting one or both ears. These marks were registered with the county clerk as underbits, overbits, crops, splits, swallowforks, and other colorful terms. These hogs were then released into the wild.

Smart hog farmers found ways to keep their hogs in the area by providing them salt, or by planting specific crops that the hogs would come to. By doing this, the farmer was able to keep an eye on his hogs, and to select some for sale or slaughter for his own needs. As the wild woods shrunk year by year, and fencing became more and more practical and eventually mandatory, the days of the hog farmer allowing his stock to run loose came to an end. Now hogs are raised under controlled conditions, from birth to marketing. No longer do sows give birth in the open fields or woods, the newborn pigs easy prey for wild animals. Pigs are born in farrowing houses, where there is climate control and the health and safety of the newborns and the mother is ensured. Hogs are fattened on scientifically formulated feeds, there is protection from insects and disease, plenty of fresh water is provided, and there is relief from the blistering Florida sun. This last point is crucial: as the pig has no sweat glands, keeping cool is a major requirement.

At birth, the weight and health of each pig is recorded.

PHOTO COURTESY OF CHAD LYONS

Pigs must be sprayed to control external parasites.
PHOTO COURTESY OF RONNIE PALMER

Many changes have come to the hog industry. The old, lard-type hog is gone. These older breeds produced a greater percentage of body fat than current commercial breeds. Lard was a valued article of trade, it was a primary fat used for all types of cooking and baking. Old estate inventories record that at times bacon and lard carried the same value, pound for pound. Changing consumer tastes, and health concerns over the intake of fat caused

Pigs are weighed at specific intervals to determine rate of gain and feed conversion.
PHOTO COURTESY OF RICKY LYONS

These gilts are fed a ration of ground corn from a homemade feed trough.

Photo courtesy of
Ronnie Palmer

Students load up a show pig selected to raise for an FFA project.

Photo courtesy of
Mark Spradley

researchers and breeders to develop breeds that were leaner, grew faster, and were gentler and easier to handle. Aggressive marketing promotion and the introduction of specialty products have caused the pork market to remain healthy. For many, barbeque means pork. Pulled pork, sliced pork, and ribs are all popular forms of barbeque. Whether it is cooked on the backyard grill, in the oven, or purchased from one of the many barbeque specialty restaurants, barbecued pork remains a top seller. The importance of pork to the barbecue restaurant industry is evidenced by the number of establishments that have "Pig" as part of their name.

Bacon, sausage, and ham remain the meat of choice for many consumers. For some of us, it just isn't breakfast without bacon or sausage.

"Living high on the hog" and being in "hog heaven" are still expressions that connote abundance, wealth, and quality of life.

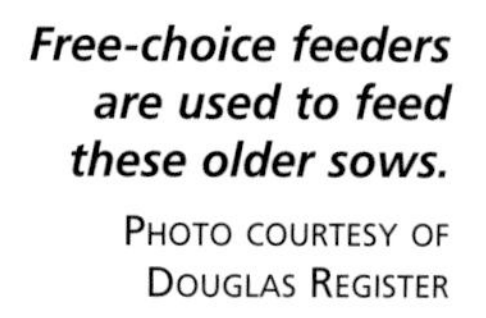

Free-choice feeders are used to feed these older sows.

PHOTO COURTESY OF DOUGLAS REGISTER

CHAPTER 7

HORSES

Paleontologists have found fossils of forerunners of the modern horse in Florida, but they were extinct long before recorded history.

Like cattle, Florida's first modern horses came with the Spanish on their early voyages of exploration. The Spanish established breeding herds of horses on some of their Caribbean island possessions beginning with Columbus, who brought horses to Haiti on his second voyage in 1493. The first horses brought to Florida came in 1521 with Juan Ponce de Leon on his second voyage. Ponce de Leon acquired these horses from the breeding herds in the Caribbean. During this period of exploration and discovery, horses strayed and were never recovered, or abandoned. These horses had to have the ability to survive the rigors of the Florida wilderness. The predators, insects, diseases, and parasites took their toll, and the number of horses in this time frame was probably never very high. The late 1500s through the 1600s

"Go Dick Go," an early Florida racing quarter horse, is shown with breeder Raymon Tucker. Both man and horse in this photo are legendary in quarter horse racing.

PHOTO COURTESY OF JOHANNA DAVIS

saw an increase in Spanish cattle ranching, and it was necessary to breed and import large numbers of horses to tend these cattle ranches. Whether it was the early strays from the Spanish, or later ones during the ranching period, there were large numbers of "wild" horses roaming free when the United States acquired Florida in 1821.

This well-mounted Florida cowboy is ready for work.

PHOTO COURTESY OF DESERET CATTLE AND CITRUS

The "cracker horse" was a small horse, thirteen to fifteen hands, rarely weighing over a thousand pounds. The term "cracker" refers not so much to the horse, as to the rider. The early Florida "cow hunters" carried a long, braided-leather whip. These "blacksnake" whips, as they are sometimes called, could be eighteen to twenty feet long, and were not used to beat the cattle, but were cracked above the heads of the cattle to drive them where needed. Cow hunters possessed great skill and accuracy with these whips. Some would demonstrate their prowess with the whip by removing the burning end of a cigarette from someone's mouth without touching anything but the cigarette. In Florida, these whips are called "cow whips," not "bullwhips" as they are known in other places. Makers of these whips are held in high esteem, and often have waiting lists of orders many months long.

The Florida "cracker horse" is a small, hardy breed, descended from the horses brought by the Spanish. They can work longer and harder than many other breeds.

Photo courtesy of Lightsey Cattle Company

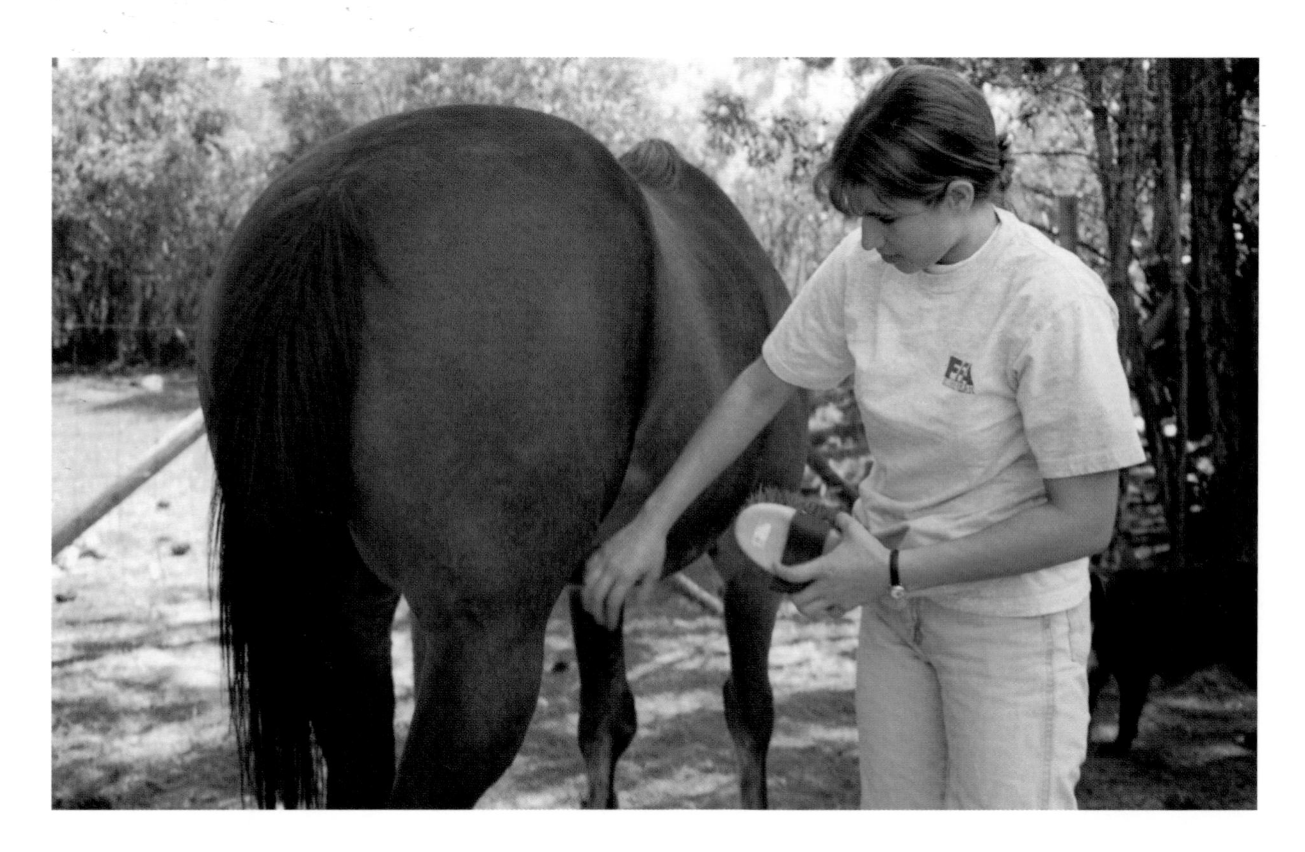

A rider grooms her quarter horse gelding before saddling up for a trail ride.

Photo courtesy of Andrea Stevenson

Cracker horses have their own registry and breed association. Herds are maintained by the State of Florida at the Florida Agricultural Museum, the Withlacoochee State Forest, and at Paynes Prairie State Park. A number of private individuals maintain breeding herds of cracker horses, and many more own one or two.

Florida's horse industry is dynamic and growing. The same conditions that make Florida a great cattle state also make it a great horse state. Horse farms in Florida produce nearly every breed of horse, from the novel miniature horse to the huge draft horse.

Florida thoroughbred breeders set out to prove that Florida horses could compete on the major racing circuit, and proved the point with finality. Florida horses have won every race imaginable. To date there have been seventeen winners of the Kentucky Derby, the Belmont Stakes, and the Preakness. In 1978 Florida-bred "Affirmed" won the golden laurel of thoroughbred racing, the Triple Crown. Florida's thoroughbred industry is centered around Marion County, one of only four major centers of thoroughbred breeding in the world.

Florida quarter horses are a major part of the horse industry also. Quarter horses have their own popular racing circuit as well. Quarter horses are the usual choice of the rodeo competitor. They are trained in roping, reining, cutting, barrel racing, bulldogging, team penning, and other aspects of the sport. These competitive events arose from the skills that were necessary to the stockman. The stockman still values a trained quarter horse and rider. Many Florida ranches could not be maintained without men and women on horseback. The quarter horse is also widely used in competitive and noncompetitive trail riding, and as "backyard" horses maintained just for the pleasure of riding. Florida quarter horses are bred, trained, and shipped all over the country for use in competition, as working horses, and as pleasure mounts.

Both horse and rider must be fearless to participate in the rodeo sport of "bulldogging," also known as "steer wrestling."
PHOTO BY MICHAEL RASTELLI, COURTESY OF WADE KAUFMAN

Nearly a quarter million Floridians are involved in the horse industry. Here a farrier is "hot shoeing."

Photo courtesy of Clayton Wilbur

There are many breeding farms in Florida, producing purebred and registered horses of nearly every breed. The horses produced on those farms, however, are probably outnumbered by unregistered, mixed breed horses and ponies that are kept for pleasure riding and amateur competition. Horses are friends and companions to many people, and are an enjoyable pastime to those who own them.

The importance of horses to Florida's agricultural economy cannot be underestimated. There are over three hundred thousand horses in Florida. Jobs related to or dependent on the horse industry number in the tens of thousands. The breeding, training, and sale of all types of horses and related businesses is a major industry. It has an estimated annual economic impact of over $6.5 billion. The popularity of horse-related activities is such that there is not a weekend in the year that you can't attend a rodeo, horse show, trail ride, field trial, fox hunt, stadium jumping, or dressage event.

Two rodeo competitors participate in team roping at the famous rodeo grounds at Kissimmee. Men and horses work together. One rider ropes the horns while the other ropes the heels.

Photo by Michael Rastelli, courtesy of Wade Kaufman

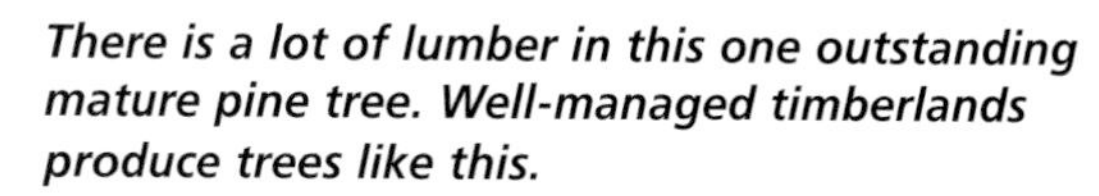

There is a lot of lumber in this one outstanding mature pine tree. Well-managed timberlands produce trees like this.

PHOTO COURTESY OF R. J. NATHE AND SONS

CHAPTER 8

FORESTRY

When the first explorers encountered Florida's woodlands, they were astounded by the trees they found there. The native Longleaf pines grew straight, true, and tall. The Live Oaks were immensely dense and tough. They instantly realized that here was the source of timber for their navies. The pines were an excellent source of masts for sailing ships, and Live Oak planking was so tough that when properly attached and braced, the cannon balls of the day would bounce off. Just as valuable as the timber was the resin of the pines, which provided the tar and pitch that waterproofed ships and protected the hulls from the dreaded shipworm that could eat through the toughest planking, destroying the ship. Florida had sufficient pines and live oaks to supply all of the navies of the world when these first explorers arrived.

The early settlers needed lumber to construct houses and towns, and timber harvest began early in the state's history. Production of lumber

Two generations of loggers head for the woods.

PHOTO COURTESY OF R. J. NATHE AND SONS

in these early days was a laborious process. Most Florida rivers gently flow to the Atlantic or the Gulf of Mexico, not providing enough fall to power water wheels to mechanize the process. No doubt these settlers appreciated every piece of lumber that had to be sawn from the log with hand-operated pit saws. In spite of this drawback, a lot of timber was cut. As each improvement in logging and milling came, there was a corresponding jump in production, peaking in the late 1880s at over one billion board feet annually. This couldn't last forever, and by the 1930s all of the virgin timber had been cut.

Reforestation programs had begun earlier, and have served to reforest the state. The Florida Division of Forestry maintains large seedling nurseries, and has the largest seed-pine orchards in the world.

Along with the lumber industry, the naval stores industry was burgeoning. There was considerable demand for turpentine and pine tar products. The native pines, and later stands of planted pines, were slashed in a descending series of shallow V's, with tin gutters nailed to the tree to funnel the crude gum into a wooden, tin, or clay container. This slashed section of the tree became known as the "cat face" and a large tree might have several on it, all being worked at the same time. Initially, the crude gum was collected and sent to ports from where it was sent overseas to be refined. Later, turpentine stills were set up in areas near where the trees were being worked, and collecting of crude gum and refining

Some big cypress logs, destined for the sawmill.

PHOTO COURTESY OF R. J. NATHE AND SONS

Timber like this pine timber on the way to the mill can be sawn into lumber, cut into veneer to be used for plywood, or peeled and treated for power poles.

PHOTO COURTESY OF KATHRYN MCINNIS

into finished products went on simultaneously. The U.S. Forest Service research station at Olustee conducted extensive experiments and disseminated information to turpentine producers, greatly increasing production and efficiency. By the late 1940s producers had nearly stopped using the open stills that had served them for so long. Sealed retorts at industrial facilities were more efficient at extracting the valuable elements from the crude gum, and most turpentine operations collected crude gum and

Modern equipment makes the job easier. In the past logs had to be skidded by teams of mules or oxen.

PHOTO COURTESY OF R. J. NATHE AND SONS

It takes a big chain saw and a skilled operator to perform the dangerous work of felling timber this large.

PHOTO COURTESY OF R. J. NATHE AND SONS

shipped it to these processors. The end of the era came in the 1950s when chemical processes for extracting gum during the papermaking process were developed, along with the retorting of stumpage from harvested pine plantations.

Florida's several varieties of pines are grown in vast stands, in all soil types, with a large percentage of them grown for the production of paper products. The Florida Division of Forestry estimated in 1995 that there were nearly seven billion cubic feet of standing pine timber growing in Florida. Pines are also grown for lumber, veneer, and plywood, with several plywood plants operating in the state. Many pines are treated with chemical preservatives and used for posts and poles. Even pine straw is collected, baled, and sold as mulch for ornamental plants.

Logs are skidded to a central point where they are loaded onto trailers to go to the mill.

PHOTO COURTESY OF R. J. NATHE AND SONS

Lumbering in other woods has always been an important part of the industry.

Millions of board feet of cypress have been cut. Cypress is a unique tree in that it grows in very wet conditions, even in standing water. The roots do not drown because the tree puts up "knees" that protrude from the water, enabling the tree to survive as long as the knees are not submerged. Cypress, a durable wood, works and finishes well. Heart cypress is durable for many years in ground contact, and cypress fence posts still are common in Florida. Cypress mulch is in great demand for use in landscaping.

Native Florida cedars were once a unique article of trade. While cedar wood is durable, beautiful, and aromatic, it was of great value to the pencilmakers. There were once sawmills at the Gulf port of Cedar Key, cutting cedar to ship to pencil factories. Teams of loggers scoured the mainland and the surrounding islands, cutting straight-grained trees, floating them back to the mills to be sawn into pencil blanks. Cedar is still a valuable wood. It is estimated that there are in excess of 120 million cubic feet of standing cedar timber in Florida now.

The magnolia is a large, attractive tree, now grown as an ornamental. It has very large, aromatic flowers, large shiny leaves, and produces cones containing bitter red seeds that are relished by the now less-than-common Pileated Woodpecker. The magnolia is symbolic of the South, and wreaths of magnolia leaves are traditional decorations on the graves of Confederate soldiers. The wood of the magnolia was once the favored wood of the basket

Machinery cuts the logs to length and puts them on a conveyer to the saw.
PHOTO COURTESY OF KATHRYN MCINNIS

Rough lumber on a conveyer going away from the saw.
PHOTO COURTESY OF KATHRYN MCINNIS

mills. Millions of trees were cut to produce baskets for shipping Florida produce.

Hardwoods of all types are found extensively throughout Florida. The estimated volume of standing hardwood timber exceeds that of the softwoods. This wood is eventually converted into lumber, veneer, and plywood.

Large numbers of trees that were cut in the nineteenth and twentieth centuries never made it to the sawmills. Some became

Finished lumber, kiln dried, planed, and ready for sale.
PHOTO COURTESY OF KATHRYN MCINNIS

A cabbage palm is harvested from the forest. It will be transplanted to another location to be used as an ornamental.

PHOTO COURTESY OF KATHRYN MCINNIS

waterlogged and sank to the bottom of Florida's lakes and rivers. Specialty companies recover these "deadhead" logs, and send them at last to their appointment with the saw. Such woods as heart pine, heart cypress, and cedar are relatively unaffected by their long submersion. These logs, when sawn, possess a trueness and beauty of grain and figure that is otherwise unobtainable today. Incomparable furniture and cabinetry have been produced from this nearly forgotten timber.

Other products are made from Florida's forests. Palm and palmetto are harvested for their fiber. Palmetto fronds are used by the Seminoles to build their traditional shelters called "chickees." The

Live oaks, draped with Spanish moss.

PHOTO COURTESY OF LIGHTSEY CATTLE COMPANY

Inset Opposite Page: Pine straw is harvested and baled and is widely used as mulch. Pine straw harvest provides income for timber owners while they wait for the trees to mature.

PHOTO COURTESY OF KATHRYN MCINNIS

cabbage palm produces "swamp cabbage," a traditional dish, as well as the hearts of palm used in salads and pickled. Palm logs are used as pilings. Palms are removed from the wild and transplanted as ornamentals. Saw palmetto berries are collected for use in medicines.

Spanish moss is one of those forest products that has seen cyclic popularity. It was once collected on a large scale, processed to remove the gray coating that covers the inner black, hair-like fiber, ginned, baled, and sent to furniture and upholstery factories for use as stuffing in furniture, mattresses, and automobile seats. Moss was replaced with other fibers for upholstery, and the collection of moss ceased. It has returned in the form of dried, preserved moss, used to line hanging plant baskets in addition to other ornamental and decorative uses, and as a substitute for sphagnum moss.

Much change has occurred in Florida's forestry industry. It remains a vital part of Florida's agricultural economy, and will remain so. Slash pines take approximately twenty years to grow large enough to be cut for pulpwood, so the trees planted today will be there for some time to come. Longleaf pine and the hardwoods take even longer. There is unlikely to be any decrease in the demand for paper, lumber, or other forest products in the foreseeable future.

This trailer load of palms is on their way to new homes.

PHOTO COURTESY OF KATHRYN MCINNIS

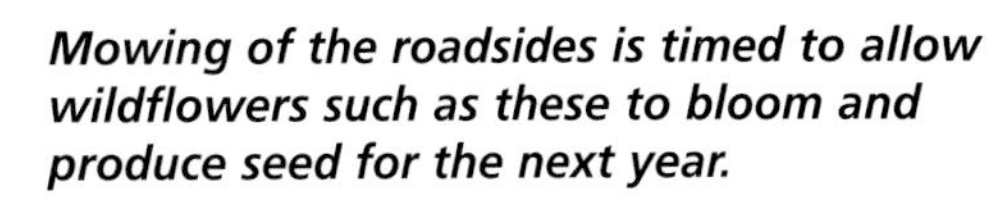

Mowing of the roadsides is timed to allow wildflowers such as these to bloom and produce seed for the next year.

PHOTO BY THE AUTHOR

CHAPTER 9

ENVIRONMENTAL HORTICULTURE

Florida truly is the land of flowers. The mild climate has something blooming every month of the year. In addition to flowering plants, cut foliage, sod, ornamental grasses, trees, and shrubs are grown as part of Florida's horticulture industry. Along with citrus and winter vegetables, ornamental plants are one of the largest of Florida's agricultural commodities. Florida is second only to California in nursery and greenhouse production. Over 85 percent of sales in the U.S. market for tropical foliage is from Florida-grown products.

The landscaping industry is the major user of these products, both in Florida and around the nation. Florida is a state of tremendous growth, large numbers of people move to the state every day. Thousands of new homes are built each year, many on land that formerly

Above: Countless acres of plants are grown in Florida greenhouses, seen here at sunset.

PHOTO COURTESY OF UF/IFAS MID-FLORIDA RESEARCH AND EDUCATION CENTER

Right: Flowering plants, such as these kalanchoe, are popular among florists and interiorscapers.

PHOTO COURTESY OF LARRY AND DEBBIE SWINDLE

consisted of fields, groves, and forests. These new developments seem to spring up almost overnight. Building codes throughout the state require varying amounts of landscaping in these new communities, creating an endless demand for sod, trees, shrubs, ornamental grasses, and annual and perennial flowering plants.

Houseplants are purchased in vast quantities by homeowners and renters. The ornamental plant industry produces plants of every variety to meet this demand. The desires of houseplant purchasers are infinitely varied. Some require plants that only need minimal care to thrive on their own. Some hobbyists insist on rare and delicate plants, needing constant attention and care to survive. The backyard greenhouse is a common fixture in Florida, and growers supply a variety of plants to fill them. Housing developments are often constructed along newly built golf courses.

Since Florida's climate is so mild, golf is a year-round game. The golfer never has to wait for the spring thaw to melt the snow

Landscapers used flowering annuals and border grasses to beautify this parking lot island.
PHOTO BY THE AUTHOR

Peace lilies are a very popular foliage plant that produce an unusual, but beautiful bloom.

PHOTO COURTESY OF GARY LEE

Mature philodendrons, in hanging baskets, are ready for market.

PHOTO COURTESY OF RAY BROWN NURSERY

Some of the twelve-hundred-plus acres of caladiums grown in Highlands County. Lake Placid in Highlands County is the "Caladium Capital of the World."

PHOTO COURTESY OF DANNY PHYPER

These six-inch to eight-inch pothos are prime specimens, ready to be shipped to willing buyers.

PHOTO COURTESY OF RAY BROWN NURSERY

Above: A greenhouse full of African violets and lipstick plants.

Photo courtesy of Larry and Debbie Swindle

Right: Ferns are grown in shade houses like the one seen here, as well as under natural shade.

Photo courtesy of Cordie Cone

before hitting the links. Turf grass for golf courses is a specialty industry developed to meet this demand. The golf greens are often the first areas to be landscaped after the land is cleared and profiled. Builders and developers learned long ago that houses would sell quicker along a completed golf course than a course under construction. Florida turf grass growers readily meet this continuing demand.

Cut foliage for the floral industry is another aspect of the buiness. While most people purchasing a floral arrangement are concerned with the flowers, these arrangements would look less appealing were it not for the ferns and other greenery that enhance the cut flowers. A large production area lies near the east coast of Florida, along the St. Johns River. Ferns and other greenery are grown in shade houses, and in natural shade.

There is increasing interest in wildflowers. Road departments are seeking ways to reduce energy costs by mowing less often, but still

These mature begonias are in full bloom. They are very popular bedding plants.

PHOTO COURTESY OF G. SCHUMACHER

The bald cypress is a native tree that is used in landscaping.

PHOTO BY THE AUTHOR

After being cut, ferns and other greenery must be kept under refrigeration at all times. Florida greenery is shipped all over the country in refrigerated trucks and planes.

PHOTO COURTESY OF CORDIE CONE

This greenhouse has been emptied of a mature crop of plants, and is being filled with flats of new young bedding plants.

PHOTO COURTESY OF LARRY AND DEBBIE SWINDLE

Crape myrtles come in several colors, and are a popular landscape tree.

PHOTO BY THE AUTHOR

Loading pallets of sod on a tractor-trailer for shipment.
PHOTO COURTESY OF LONG AND SCOTT FARMS

want the roadways to look attractive. Wildflowers offer the advantages of being easy to grow, many types will reseed themselves, fertility and irrigation requirements are low, and there are usually not a lot of pests attracted to wildflowers. Experimental plantings of wildflowers have been made, and deemed to be successful. Seminars to educate potential producers, seed processors and other interested parties to this new phase of agriculture are held, with considerable interest being shown. As this program expands, Florida's roadways will become more attractive, without additional expense.

The sod-cutting machine cuts the sod and brings it up a conveyor, where it is stacked on a pallet.
PHOTO COURTESY OF HOLT FARMS

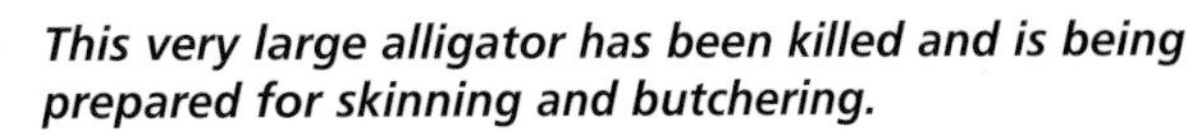

This very large alligator has been killed and is being prepared for skinning and butchering.

PHOTO COURTESY OF CLABROOK FARMS

CHAPTER 10 AQUACULTURE

One look at a map indicates that water is a big portion of the Florida environment, as it is bordered on the east by the Atlantic Ocean, with the warm Gulf Stream just offshore, on the west by the Gulf of Mexico, laced in every direction by rivers and streams, and dotted with countless lakes, large and small. The generally abundant rainfall keeps the lakes and rivers recharged.

The native fish and shellfish, both from fresh and salt water, have provided a steady supply of food for generations. The Native Americans and the early settlers relied on this harvest from nature. Many of the Native Americans lived along the coastlines, and early settlements were in coastal areas. Increasing population, loss of habitat, environmental concerns, degradation of water quality, and other factors have limited this resource. Rising to the challenge, Florida aquaculture producers have innovated, experimented, and found new methods of production.

These young alligators, seen here in a hothouse, are fed specially formulated feed, and are cared for like any other livestock.

PHOTO COURTESY OF CLABROOK FARMS

Florida aquaculture contributes over $186 million annually to total agricultural sales. Tropical fish, aquatic plants, clams, shrimp, oysters, alligators, and food, sport, and game fish are all produced in Florida.

Tropical fish production is the largest element of Florida aquaculture. Sales of these ornamental fish make up 43 percent of the total industry. Ninety-five percent of all tropical fish sold in the United States are grown in Florida. Guppies, swordtails, angels,

A hundred alligator hatchlings are ready to go into a hothouse to be raised to harvestable size.

PHOTO COURTESY OF CLABROOK FARMS

Alligators ready for skinning. Farm-raised alligator skins have replaced the once lively trade in wild (and illegal) alligator skins.

PHOTO COURTESY OF CLABROOK FARMS

tetras, gouramies, koi, mollies, and tropical catfish are only a few of the varieties of colorful and exotic fish that grace aquariums in many homes. Over eight hundred varieties of tropical fish are produced, with sales exceeding $44 million.

Aquatic plants make up another significant part of Florida aquaculture, representing about 21 percent of the industry. Plants are grown for aquariums, for water gardens, and for use in the creation of new wetlands and the remediation and restoration of

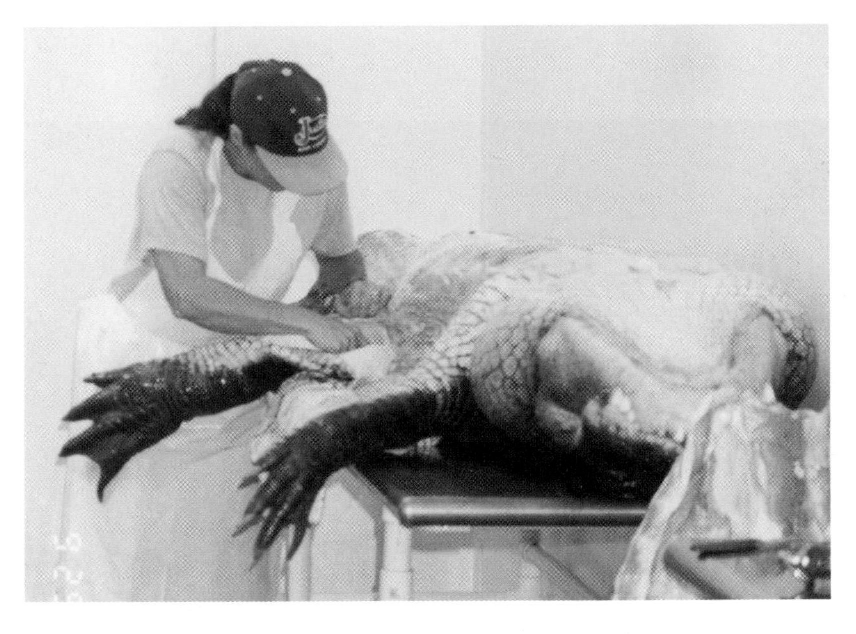

Alligators, such as this very large one, must be carefully skinned and butchered. Careless skinning can turn a very valuable hide into nearly worthless scraps.

PHOTO COURTESY OF CLABROOK FARMS

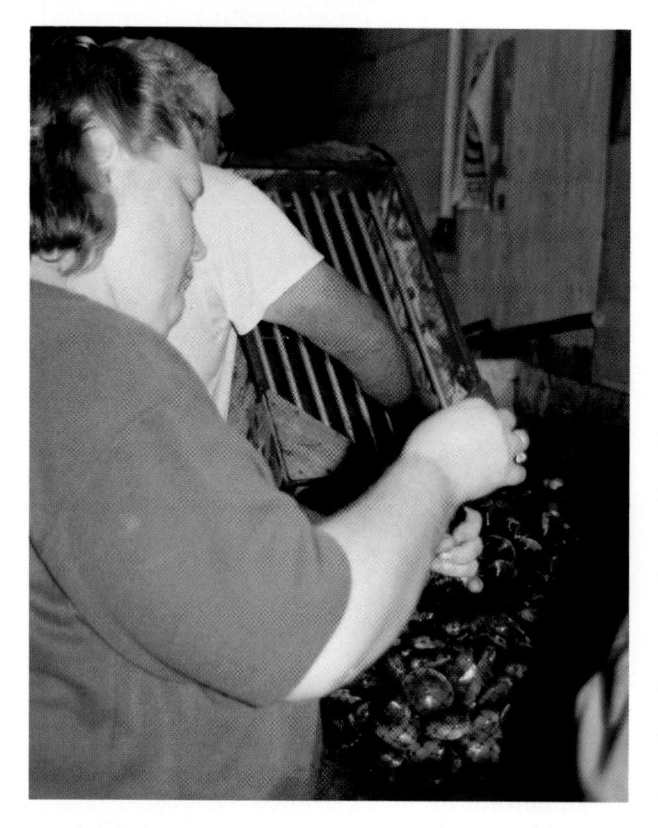

Bagging graded clams for sale.

Photo courtesy of Kathryn McInnis

existing ones. Aquatic plants help filter and purify surface water before it enters the aquifer. Alligators are big business in Florida. Tourists coming here like to see alligators, at a distance, and alligator parks and shows are popular attractions. The sale of alligator products, mostly hides, tops $3 million annually. Alligator tail is a staple of cracker cookouts, and is offered by some restaurants.

A constitutionally mandated ban on inshore netting on both the Atlantic and Gulf Coasts ended a way of life for many fishermen. But where one door closes, another opens. Many of these former net fishermen are now clam farmers. The state government wisely stepped in with financial and educational assistance to train and equip the producers for this new undertaking. Farm-raised clams now make up over 18 percent of the aquaculture industry. Oysters are also grown on managed, leased sections of sea bottom. Sales of oysters and clams are over $16 million annually.

Annual sales of catfish are over $1.3 million, and the number of catfish producers continues to increase.

Clams are sized and sorted. A sizable industry of "farm-raised" clams has sprung up along the Big Bend area of Florida's Gulf Coast. When net fishing was ended by law, many fishermen took up clam raising.

Photo courtesy of Kathryn McInnis

CHAPTER 11 POULTRY

"A chicken in every pot" was one of the great campaign slogans in a bygone presidential election year. This was during the economic quagmire of the Great Depression, when almost anything in every pot would have been welcome. It expressed what is still a great American tradition, the consumption of chicken. For many people, it wouldn't be Sunday dinner without fried chicken, and the more than 115 million broilers produced in Florida indicate that a lot of people are eating chicken. Particularly when one considers that only about 50 percent of the state's consumption of chicken is produced here, and the remainder must be imported from other states.

Chicken finds its way to the table in many forms. Recent years have seen an upswing in what would have been nontraditional cuts just a few years ago. Boneless breasts, filets, tenderloins, and nuggets are all portion-controlled cuts that the consumer is quite fond of. Franchised fast-food

Chicken houses are a common sight on North Florida farms. The climate and proximity of processing plants make this area suitable for poultry production.

PHOTO COURTESY OF SHIRLEY CARTE

restaurants serve up a lot of this type of chicken in addition to the traditional "grab and go" burgers. The popularity of chicken wings is phenomenal. Once a less than popular cut of chicken, wings are much in demand. Most pizza restaurants carry wings in addition to the traditional Italian dishes they are known for. The

Broilers are ready for market when they are several weeks old.

PHOTO COURTESY OF SHIRLEY CARTE

Young chicks are delicate and sensitive to temperature changes. Heat lamps are positioned over them to keep them warm.

PHOTO COURTESY OF SHIRLEY CARTE

patrons of "sports bars" consume vast quantities of wings. The "Monday Night Football All-You-Can-Eat Wing Special" is stock-in-trade for these businesses. You can have them barbequed, mild, spicy, really spicy, Cajun, teriyaki, and other flavors.

Florida broilers are marketed from four weeks of age for "Cornish Hens," up to ten weeks of age for full-grown broilers. The 115-million-plus broilers produced have a production value of over $250 million. Producers give a specially formulated feed to their flocks, determined by the best research available. Poultry production is very intensive, hands-on agriculture. Florida's heat and humidity make it a necessity to have cooling fans and other methods of keeping the temperature regulated to prevent a high rate of mortality. A major power failure or a prolonged heat wave can spell disaster for the poultry producer. On the other hand, chickens, particularly young chickens, are sensitive to cold, so in

There's a lot of chicken feed in that bin. Broiler feed is made up to specific formulas by the processors.

PHOTO COURTESY OF SHIRLEY CARTE

the cooler months it is necessary to provide heat lamps and curtains for the chicken houses so that the temperature does not fall too swiftly or too far.

Florida egg producers turn out nearly 3 billion eggs per year from a total flock of nearly 11 million hens. The value of Florida's egg production is over $120 million. Eggs, like chicken, have found their way into the fast-food markets. Breakfast sandwiches utilizing eggs are available at virtually all of the franchised fast-food restaurants. These sandwiches are quite popular, and probably more people are consuming eggs in this manner than in the traditional sit-down breakfast.

The production value of Florida's poultry industry is in the area of $380 million annually, and that's not "chicken feed" by anyone's standards.

CHAPTER 12 TOMATOES

The importance of tomatoes to Florida's farm income cannot be ignored. Thirty-five percent of the sales volume of Florida vegetables is tomatoes. The value of the tomato crop has reached nearly $600 million in recent years. Florida is the largest producer of tomatoes in the nation. During the winter months, Florida supplies nearly 100 percent of the fresh tomatoes throughout the country. Tomatoes are grown in nearly every part of the state. Florida is divided into five growing areas or markets. The Immokalee market produces two crops a year, with a crop harvested in March and April. The Palmetto-Ruskin and East Coast markets also produce two crops a year, harvesting the early crop in April, May, and June, and the later crop in October, November, and December. The Palmetto-Ruskin market produces more

Juicy, ripe tomatoes. Florida growers produce more tomatoes than any other state. Almost 100 percent of the fresh tomatoes in the United States during the winter come from Florida.

PHOTO COURTESY OF THOMAS WRIGHT

tomatoes than any of the other districts. The Homestead market harvests one crop each year during December, January, and February. The northern production area produces two crops a year, one in June, and another in October and November. The northern area has a shorter growing season than the other four areas because of climate.

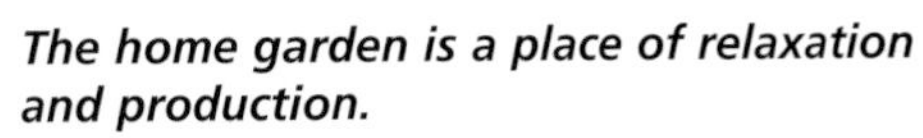

The home garden is a place of relaxation and production.

Photo courtesy of Douglas Register

Chapter 13 Vegetables

Florida vegetables are grown throughout the state, with planting and harvest times moving up and down the state with the seasons. The state is divided into eight vegetable-producing areas, each having some specialty that it does best. Some of these regions are known for the crops they produce. Ruskin tomatoes are widely known, and the name Zellwood is almost synonymous with sweet corn. Hastings is known for potatoes and cabbage. There is much overlap in these areas. For instance, the West Florida area produces fine crops of tomatoes and pole beans, as does the Southeast area at the other end of the state. These widely separated areas produce their crops on different schedules, due to the difference in climate. Much of the Lower Peninsula is nearly frost free during the winter, and crops can be produced there while the rest of the country remains locked in the icy grip of winter. As spring comes to Florida before the other states, planting and hence

Multiple plantings can be used to supply the needs of the family or the market. Careful cultivation is necessary to keep out weeds.

PHOTO COURTESY OF DOUGLAS REGISTER

This cabbage is ready for harvest. Hastings, in St. Johns County, is famous for cabbage and potatoes.

PHOTO COURTESY OF DOUGLAS REGISTER

When harvesting and packing cabbage in the field, heads that are not acceptable are left behind.

PHOTO COURTESY OF DOUGLAS REGISTER

This broccoli is ready for harvest. It is usually grown as a cool weather crop in Florida.

PHOTO COURTESY OF DOUGLAS REGISTER

Washing and packing cucumbers.
Photo courtesy of
Frankie Allen

marketing precedes the rest of the nation. Often there is a crop available in the store before other states can even get their crops planted.

The major vegetables produced include snap beans, cabbage, sweet corn, cucumbers, eggplant, bell peppers, radishes, squash, tomatoes, and potatoes. There is also significant production of butter beans, field peas, greens, cauliflower, cherry and plum tomatoes, lettuce, escarole, celery, hot peppers, and okra. Innovation and experimentation continue in vegetable production: herbs are grown both in greenhouses and open fields, and carrots are being produced in North Florida, where there has never been commercial production. There are numerous other

Sweet potatoes are grown from vegetative cuttings that have been rooted. Here the cuttings or "draws" are being sorted and separated before planting.

PHOTO COURTESY OF STARLING FARMS

Above: Boxes of sweet potatoes are dried and stored in bulk in tobacco curing barns prior to marketing.

Photo courtesy of Starling Farms

Right: A popular green in the South, collards can be harvested several times during the season.

Photo courtesy of W. A. Fish

Sweet corn must be tilled and sprayed regularly to control weeds and insects.

PHOTO COURTESY OF
FRANKIE ALLEN

Collard greens are packed in boxes using a conveyor belt. The packed boxes will be kept under refrigeration.

PHOTO COURTESY OF
W. A. FISH

These crates of sweet corn are ready for shipment in a refrigerated truck.

PHOTO COURTESY OF FRANKIE ALLEN

vegetables produced, some for specialty markets, including ethnic and cultural groups. Florida's growing international community fuels demand for vegetables that are familiar in other countries, but are unknown here, and Florida farmers have risen to the challenge to supply this market. As you can see, the market basket is full with an endless variety.

Sweet corn is harvested and packed in the field on mobile harvesters.

PHOTO COURTESY OF FRANKIE ALLEN

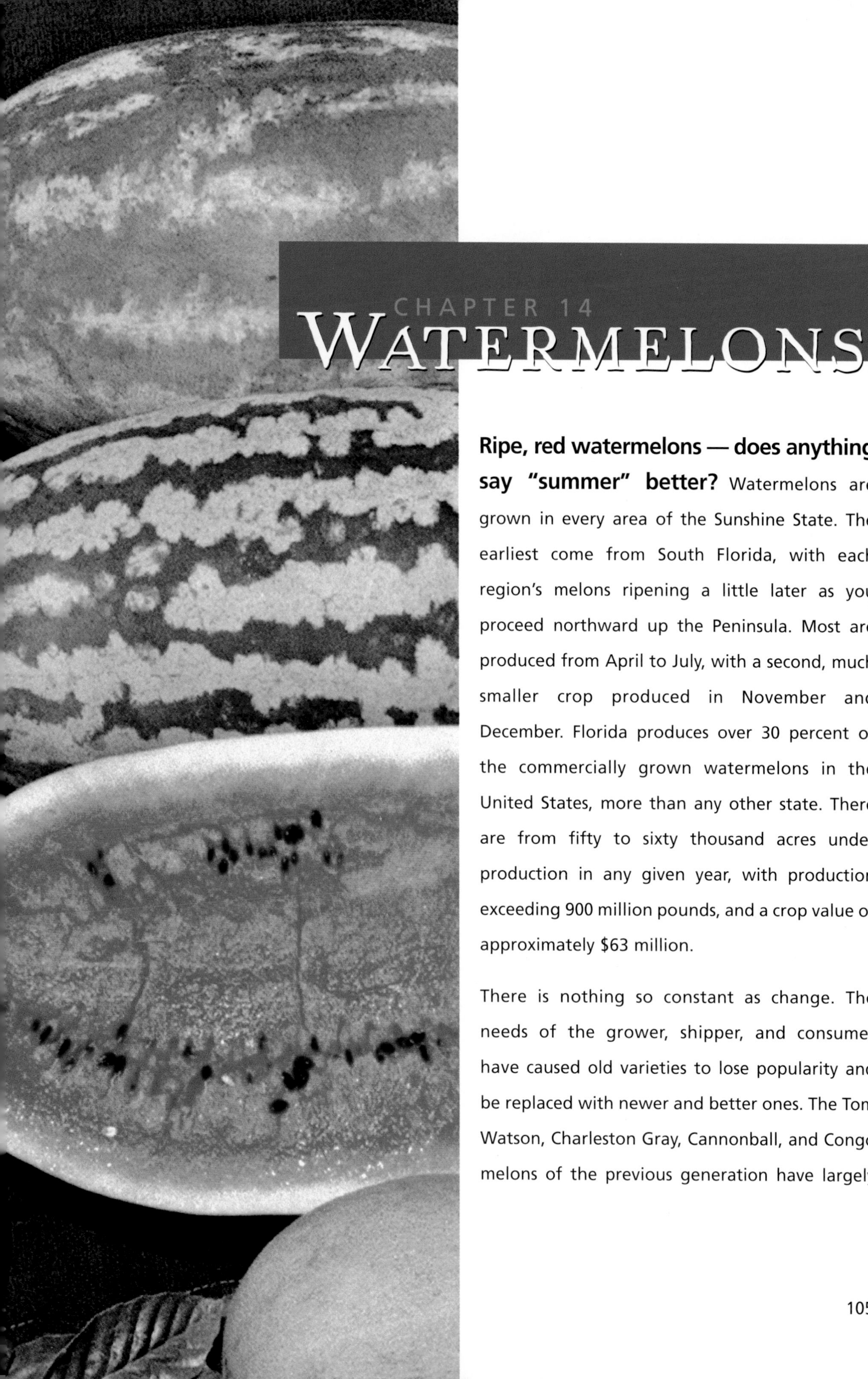

CHAPTER 14

WATERMELONS

Ripe, red watermelons — does anything say "summer" better? Watermelons are grown in every area of the Sunshine State. The earliest come from South Florida, with each region's melons ripening a little later as you proceed northward up the Peninsula. Most are produced from April to July, with a second, much smaller crop produced in November and December. Florida produces over 30 percent of the commercially grown watermelons in the United States, more than any other state. There are from fifty to sixty thousand acres under production in any given year, with production exceeding 900 million pounds, and a crop value of approximately $63 million.

There is nothing so constant as change. The needs of the grower, shipper, and consumer have caused old varieties to lose popularity and be replaced with newer and better ones. The Tom Watson, Charleston Gray, Cannonball, and Congo melons of the previous generation have largely

A sweet, juicy, field-ripened Florida watermelon. An example of one of the increasingly popular seedless varieties.
PHOTO COURTESY OF THOMAS WRIGHT

been replaced with a dozen or more hybrid varieties with superior disease resistance, shipping and marketing qualities, and productivity.

Seedless watermelons have been grown in the United States for more than forty years, but it has always been a difficult and expensive crop to grow. The wide availability of seeded watermelons at low cost had always kept the seedless melon from being very competitive. Research and the subsequent development of better varieties along with aggressive marketing have all come together to make the seedless melon a viable competitor in the supermarket. Producers have a dozen or more seedless varieties to choose from also.

Large, round, or oval melons still predominate the market, but there is increasing interest each year by consumers for the smaller, "icebox" type melons. Watermelons are available in the familiar red-fleshed varieties, with smaller numbers of yellow- or orange-fleshed varieties produced for specialty markets.

Watermelons are one of those crops that must be fully ripened on the vine in order to have acceptable quality. Unlike some fruit, which will continue to ripen after it is picked, the watermelon will not develop any further sweetness once it is removed from the vine. The vines must be kept healthy in order to mature the crop, and skilled workers are necessary to select only fully ripe melons for harvest.

Some communities recognize the value of watermelons to the local economy, and places like Newberry in Alachua County and Chiefland in Levy County hold annual festivals to celebrate the crop.

CHAPTER 15
STRAWBERRIES

Sweet, juicy, red strawberries are delightful to look at as well as eat. Cultivation of the strawberry has come a long way, from the small, but tasty, wild ancestor of our modern berry, to the large, well-developed fruit that we see in our stores today. There are four major production areas for Florida strawberries. The northern area centered around Bradford and Union Counties, the southeastern crop in the Homestead area, the Palmetto-Ruskin area, and the justly famous Plant City area.

Strawberries are one of those crops that still must be harvested by hand. No mechanical strawberry picker exists, nor is one ever likely to. The fruit is very delicate, ripening one at a time, rather than all at once.

Florida growers provide over 95 percent of the strawberries to the U.S. market in the winter. Variety development by the researchers at the University of Florida's Strawberry Research Center

Quite a handful! Any grower would be proud of these red ripe strawberries. They must be picked ripe, as they will not ripen after picking.

PHOTO COURTESY OF ILA ALLEN

U-pick strawberry fields are a popular option for the strawberry lover. Outings to the strawberry fields are often family affairs.

PHOTO COURTESY OF LARRY AND DEBBIE SWINDLE

Strawberries and cream is one of the countless ways to enjoy them.
PHOTO COURTESY OF ILA ALLEN

has introduced several improved cultivars designed to grow and produce in the unique climate of Florida. Their successes have created a phenomenal industry. Growers produce an average of around 15 million twelve-pound flats of strawberries in a production season that is only five months long. The average value per flat is nearly $12, making the total crop value over $167 million.

By far, the Plant City production area is the most well known. The annual Strawberry Festival brings hundreds of thousands of visitors to the city. Nationally known entertainers perform as well as local talent. Strawberries are available in many forms, and several hundred thousand strawberry shortcakes are sold at the event. A trip to the Plant City Farmers Market during the peak of the season will find the loading docks bustling with activity. Strawberries must be handled quickly and carefully if they are to make their long journey to the North safely. Savvy consumers can

Everybody likes strawberry and chocolate pie!
Photo courtesy of Ila Allen

also be found at the Farmers Market, haggling with their favorite vendor for berries to take home and turn into jam, freeze, or eat fresh. Chartered busses arrive at strawberry stands around town, and flats of berries go into the luggage bays to go home with the passengers.

The popularity of the strawberry has shown no signs of decline. The market continues to expand, and improved varieties and production methods will ensure that the strawberry will have better flavor, longer shelf life, and will be available in quantity.

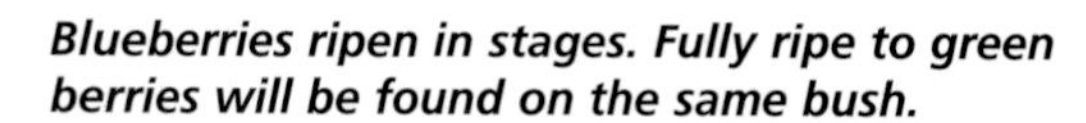

Blueberries ripen in stages. Fully ripe to green berries will be found on the same bush.
PHOTO BY THE AUTHOR

CHAPTER 16

BLUEBERRIES

Wild, native blueberries have always grown in the forests of Florida, a fact well known to the Native Americans, the wildlife, and the crackers that settled the interior of the state. The acid flatwoods soils are particularly suited to blueberries. These native blueberries are small and relatively unproductive. Through research and development, varieties have been introduced that are large, attractive, flavorful, and highly productive. Mild winters allow Florida blueberries to bloom and set fruit earlier than those in neighboring states to the north, allowing Florida blueberries to reach the market ahead of other states.

Two types of blueberries are generally grown in Florida: rabbit-eye blueberries and southern highbush blueberries. The rabbit-eye once was the predominant type planted, but more growers are planting southern highbush types. On average, southern highbush blueberries ripen about a month earlier than rabbit-eye types and

Fully ripe blueberries. Prime fruit like this is consumed fresh, made into pies and jam and countless other recipes. You just can't go wrong with fruit this good.

PHOTO COURTESY OF THOMAS WRIGHT

also produce better as plantings move down the state. U-pick operations tend to plant both types in order to extend the production season.

North Florida and West Central Florida areas produce the bulk of the blueberry crop. Areas farther south in the state are developing into production areas as a demand for early season blueberries expands. Variety development and improved agricultural practices have proven that blueberries can be produced in many areas of the state. Some of the same growing conditions required for strawberries are also ideal for blueberries. Cash receipts are in excess of $11 million.

Most Florida blueberries are shipped north for the fresh fruit market. There generally is an excellent market with good prices if the berries can be shipped before the competing states' berries come on the market. Some Florida blueberries go for local consumption. U-pick operations are also popular in some areas. Very little of Florida's production goes for processing.

A number of health benefits are claimed for blueberries. They contain dietary fiber, antioxidants, vitamin C, and other beneficial components. They contain few calories and almost no fat. They taste good too.

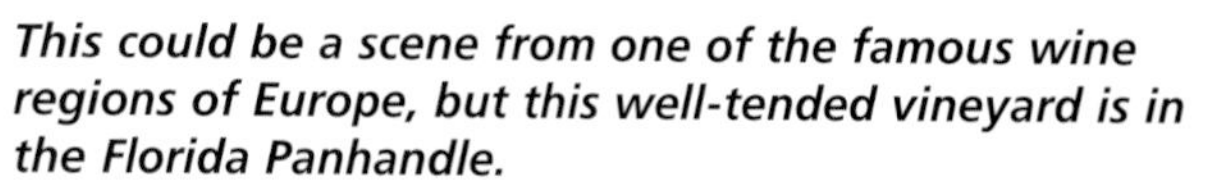

This could be a scene from one of the famous wine regions of Europe, but this well-tended vineyard is in the Florida Panhandle.

PHOTO COURTESY OF THOMAS WRIGHT

CHAPTER 17

GRAPES

The early European settlers brought grapevines with them, varieties that they had grown successfully in the Old World and that they hoped to transplant to the fertile soil of Florida. Initially, their efforts met with success, and traditional bunch grapes were grown here. Ultimately, their dreams of vineyards were not realized, because Florida soil contained pests and diseases lethal to European grapes. Pierces' disease was the death knell for bunch grapes for many years. Like other "failed" crops, the bunch grape has refused to die. Ongoing research since 1945 has developed several varieties of traditional bunch grapes, disease resistant, with good market characteristics. The dreams of the first grape growers may yet come to pass.

There has always been another option for those desiring to grow grapes in Florida. The native muscadine grape has been traditionally grown as dooryard fruit on overhead trellises or arbors. For generations before they were discovered commercially, these grapes supplied cracker families with a source of fresh fruit, and fruit for

processing into jelly. The grape arbor also provided a cool, shady retreat on hot summer days. These grapes differ from the familiar bunch grape, in that while some varieties may grow in clusters, each grape ripens individually. Muscadines are easy to grow and very few diseases or pests afflict them. Muscadines have been bred and developed into numerous varieties, with different ripening periods, colors, flavors, and intended uses. Some muscadines are best for wine, others for juice, still others for jelly, but much of the crop is consumed as fresh fruit.

A bumper crop of Florida muscadine grapes. This is one of the purple varieties available to growers and homeowners. Florida vineyards are popular u-pick destinations for those wanting the freshest grapes possible.

Photo courtesy of Thomas Wright

Many muscadines are grown for the u-pick market. When they ripen, vineyards open their gates for families and individuals to select their favorite varieties, pick them, and process them for whatever use they intend. Some vineyards have presses, and for a nominal fee will press the fresh grapes to extract the juice. The muscadine has its own unique flavor, unlike that of the traditional grape. The grapes themselves have a tough, inedible skin, which surrounds a very juicy pulp. Muscadines do have seeds, but research is ongoing to find and develop seedless varieties, as well as those with thinner, more edible skins. Muscadines have found their way into grocery stores, and consumers willing to try them find a unique, enjoyable product, uncompromisingly American.

Commercial vineyards and wineries have come into being in recent years. With proper variety selection, intense management, irrigation, and appropriate cultural practices, these wine grapes are very productive. The wines produced are unique, very flavorful, and are finding increasing popularity. Both bunch grapes and muscadines are used for wine production. Florida commercial wineries can expect yields of five or more tons of grapes per acre on well-managed vineyards. Approximately 160 gallons of wine can be produced from a ton of grapes.

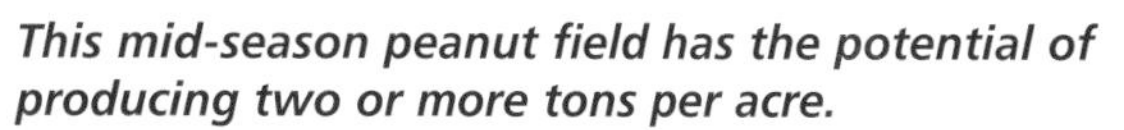

This mid-season peanut field has the potential of producing two or more tons per acre.

PHOTO COURTESY OF RALPH YODER

CHAPTER 18

PEANUTS

The most popular nut in the country isn't a nut at all, and it doesn't grow on a tree. The peanut is a legume, a relative of the bean, and is unique in that after the flower is fertilized, the pod begins life above ground, then burrows into the soil and develops underground. The peanut is a native of South America, where it was highly regarded. There are representations of peanuts made of gold that have been found in ancient burial sites, and remains of peanuts have been found in archeological digs.

The work of the great agriculturalist George Washington Carver brought much-deserved attention to the peanut. Dr. Carver developed and demonstrated many products that could be made from peanuts.

With the demise of cotton in many areas because of the devastation of the boll weevil, the lowly peanut became the savior for many farmers. Much of this cotton land was "worn out," the

Early season peanuts in bloom.
Photo courtesy of
Ralph Yoder

fertility depleted through the continual cropping of cotton. Because the peanut is a legume, it "fixes" or traps nitrogen from the atmosphere through specialized cellular structures in the roots, enriching the soil as these roots break down after the life cycle of the plant is finished. Peanuts provided food for people and animals. It was once common to plant large tracts of peanuts to

Depending on the weather, mature peanut plants will dry for two or three days before the combine arrives for the harvest.
Photo courtesy of
Ralph Yoder

The peanut plow uproots the mature peanut plants and inverts them, leaving the peanuts that were formerly underground above ground so that they can dry before harvest.

Photo courtesy of Ralph Yoder

fatten hogs on. The expression "root hog or die" is easily applied to this practice. The hogs had to dig up, or root, the peanuts out of the ground in order to eat them. The natural oils and the high-protein content of peanuts fattened hogs at a rapid rate. Peanut-fed pork is almost impossible to find today. While it is very flavorful, it does not get firm when chilled and is unacceptable to meat processors because of this characteristic.

The peanut provides the favorite snack food of the South, the boiled peanut. Immature or "green" peanuts are boiled in salted water until tender. Football games, races, rodeos, and other outdoor events just wouldn't be the same without boiled peanuts. Green peanuts freeze well, so they can be enjoyed at any time of the year. They can be found in cans in some grocery stores, and the roadside vendor of boiled peanuts is a common sight. Peanuts can be shelled and unshelled, salted and unsalted, honey roasted, chocolate coated, in candy, in cookies, the list goes on and on. Peanut oil is excellent cooking oil because it has a good flavor and high-

After these modern multirow peanut combines do their work, peanuts will be graded and dried before they are shelled and processed.

PHOTO COURTESY OF RALPH YODER

temperature tolerance. And who can ignore peanut butter? There is hardly a home in America that doesn't have a jar in the pantry.

Through research and development, the peanut has become much more productive. Proper selection of variety and scientific management have caused yields to exceed two tons per acre for some growers. When the peanuts mature, the vines are dug from the ground with specialized plows that invert the plants. The peanuts that grew underground are now exposed to the sun and allowed to dry. When they have dried sufficiently, a combine picks them up, and separates the peanuts from the vines. The peanuts are transferred from the combine into a dryer wagon where they are dried by forced air to prevent them from spoiling. They then proceed to the crusher if they are to be used for oil, or to the sheller and roaster as they continue on their journey to becoming the food products we are familiar with.

Oats, shown here in mid-season, are also grown for winter grazing as well as grain.
PHOTO COURTESY OF RALPH YODER

CHAPTER 19

FORAGE

Forage crops include grass for hay; rye, ryegrass, oats, perennial peanuts, clover, corn, and sorghum for silage; and other crops produced for grazing, for hay, and for silage. In former years, forage crops on Florida farms were dominated by fodder, which consisted of corn leaves that were pulled by hand, dried in the sun, and stored for winterfeed. This practice was found to be detrimental to the corn crop, and didn't provide that much nutrition for livestock. It was also hot, backbreaking labor. Fodder had to be pulled while the corn was at the peak of its growth, which is also when the summer heat is the worst. The almost daily rains that sweep across the Peninsula during the summer also dictated that the fodder had to be hastily put under shelter when a rainstorm approached, and put back in the sun to dry after the rain had passed.

Peavine hay was also a staple forage crop. Peas were a better source of nutrition than fodder. A

Cattle egrets follow many farming operations, such as mowing hay, consuming the insects that the tractor and equipment stir up.

Photo courtesy of Charlene Beebe

common variety was the iron clay pea, which produced a prolific growth of vines, particularly if nitrogen fertilizer was added. While peas made excellent hay, it was still hard work. The heyday of peavine hay was before the advent of the efficient and laborsaving haymaking equipment that is available to the farmer today. It also

This field of wheat in the Florida Panhandle is ready for harvest. While Florida doesn't produce wheat like the Midwestern states, some varieties do well. Wheat is also planted for winter grazing.

Photo courtesy of Ralph Yoder

High-quality coastal Bermuda hay is rolled tight, ready for winter feeding.
PHOTO COURTESY OF CHARLENE BEEBE

took more time to dry than grass hay, thus increasing the danger of a rainstorm soaking it and lowering the quality.

Corn and velvet beans were once grown together in the same row to feed livestock. The velvet bean is a large bean, somewhat resembling a lima bean, that is contained in a pod that has a soft, velvety coating on it, thus giving it its name. The velvet beans would use the cornstalks for support, climbing up them to the tops of the stalks. Cattle and hogs would be turned into fields of corn and velvet beans to feed. The stock had to work for this feed, they had to tear down the cornstalks with their drapery of velvet beans before they got to eat. Velvet beans are but a memory to many older farmers, and a quaint story about the old days to the younger generation, but once in a while you find a farmer who saves a few seeds, just to keep them from dying out.

Other crops have come and gone. Chufas, a plant that produces small underground tubers, were once widely planted for hogs to

Corn is chopped for silage while still green. The silage process ferments it and produces a high-quality cattle feed.

PHOTO COURTESY OF SHIRLEY CARTE

"root out." Chufas are now planted for wildlife management. Wild hogs and turkeys are fond of them. Peanuts were also planted for hogs to forage on. And while peanut-fed pork is far more flavorful than corn-fed pork, the meat did not get firm when chilled, which made it difficult for the meatpackers to handle. Demands for leaner pork also sounded the death knell for peanut-fed pork.

Forage production today is dominated by grass for grazing and hay, and by corn for silage.

Grass hay is widely produced in all parts of the state. It is well adapted to mechanical harvest and handling. Grass is baled in large, medium, and small round bales, large square bales, and the traditional small square bales. Round balers can wrap the bales in plastic sheeting making them weatherproof, reducing the cost to the producer for storage.

Some legumes are also produced for hay. Perennial peanut hay is a good source of nutrition, and has been compared to alfalfa hay. Horse producers have perennial peanuts as an alternative to alfalfa hay, which must be trucked in from the Western states, and sometimes contains blister beetles, which are deadly to horses. Some clovers do well enough in the Panhandle to be used for hay. Soybeans have been cut and baled for hay, and there might even be somebody out there still baling a few peavines.

Silage is a staple feed for the cattle industry. The dairy industry would have a hard time surviving without it. The process of making silage is one of controlled fermentation. In the usual process, the crop is chopped by harvesters specifically designed to

A bunker silo is being filled and packed. The silage must be packed airtight, otherwise it will rot rather than ferment.

PHOTO COURTESY OF SUWANNEE RIVER R C & D

cut the plant material to the optimal size for proper fermentation. The raw silage is then placed in an upright silo, a trench silo, a bunker silo, or in specially designed plastic bags. The raw silage must be packed and sealed to exclude air. If the seal is broken while the silage is fermenting, the silage will rot rather than ferment. When the silage has cured, this high-protein and obviously well-liked feed is fed to both beef and dairy cattle. Corn and sorghum are the most common crops used for silage. The demand for silage is large enough that there are specialized contractors cutting, packing, and sealing silage for farmers.

Modern equipment rolls the hay, wraps it with twine and/or plastic sheeting, and ejects the roll onto the ground.

PHOTO COURTESY OF CHARLENE BEEBE

Other crops can also be turned into silage. Small grains and grasses are cut and baled into large round bales while still green. The bales are then sealed in large plastic bags and left to undergo fermentation.

Sorghum is another crop that is readily chopped and converted into high-quality silage.

PHOTO BY THE AUTHOR

These tobacco plants are only about three months old.

COURTESY OF SHIRLEY CARTE

CHAPTER 20

TOBACCO

Tobacco is another of those crops that is native to the Americas, and was used ceremonially by the indigenous people for centuries before the first European explorers arrived. Tobacco was grown extensively by the early colonists, and shipped to Europe in large quantities. Tobacco was processed into snuff, smoking tobacco for pipes, cigars, chewing tobacco, and cigarettes.

Florida was found to be a hospitable climate for tobacco, even though the long growing season also ensured a plentiful supply of insects. Until the advent of chemical and biological controls, the tobacco producer had to carefully examine each plant and remove any insects by hand.

Tobacco has always been a labor-intensive crop. Innovations in production methods and machinery have removed much of the hand labor, but much still remains. Old-timers will tell you that it took nearly a full year to prepare for the crop, produce and market the crop, and destroy

Tobacco seeds are very small and very expensive. Specialized equipment is used to plant them in seedbeds in January. The seedbeds will be protected with plastic sheeting until all danger of frost is past.

Photo courtesy of Shirley Carte

the residue of the crop to prevent the overwintering of another generation of pests. In the days before chemical fumigants were introduced, preparation began in the late summer or early fall, with the cutting of brush and tree limbs to be dried and used as fuel for the fires that sterilized the soil where the seedbeds would be planted. The seedbeds, like the crop, could not be planted on

Mechanical transplanters are used to set the plants in the rows uniformly and rapidly, but each plant must be handled individually.

Photo courtesy of Starling Farms

Young tobacco plants are pulled from seedbeds by hand in preparation for planting in the open fields.

PHOTO COURTESY OF STARLING FARMS

the same land in successive years because of soil-borne diseases and nematodes.

The seedbeds were planted early in the year and were covered with cheesecloth, or in later years, clear plastic, to protect the seedlings. The young tobacco plants are easily frozen, and many a tobacco grower has had his seedbeds wiped out by late spring frosts and freezes. When (hopefully) the danger of freezing temperatures is past, and the plants are several inches high, they are pulled from the seedbeds and planted into rows in fields that have undergone preparation. In the early days the plants were set out one at a time, often with a sharpened stick used as a planting tool. The hand transplanter was invented, which allowed a worker to dig a hole, insert a plant, and release a small quantity of water

Tillage is necessary to control weeds which rob the developing plants of nutrients and lower the quality of the crop. Weeds also make harvesting difficult.

PHOTO COURTESY OF STARLING FARMS

Irrigation is necessary during dry periods. It is important that the crop is not stressed and grows uniformly, which creates a higher quality product.

PHOTO COURTESY OF STARLING FARMS

A mechanical topper is used to remove the flower from the tobacco plant. This prevents the plant from producing seed, which would lower the quality and yield. This process was once done exclusively by hand.

PHOTO COURTESY OF CLIFF STARLING

Early tobacco harvesters used workers that rode on the machine as it was pulled through the field. The "croppers" as they were called, still had to break the leaves from the plant by hand, but they didn't have to walk all day.

PHOTO COURTESY OF DOUGLAS REGISTER

Tobacco was readied for the curing barn on the overhead platform of this early harvester.

Photo courtesy of Starling Farms

into the hole with the plant. Later still, horse-drawn and tractor-drawn transplanters allowed workers to ride while the machine opened the furrow, injected water into the rooting zone, and closed the furrow after the workers had inserted the plants. The fields had to be fertilized and kept weed free. The plants had to be continually monitored for insects and disease, and the blooms and any suckers growing on them had to be removed by hand.

Harvesting was done by hand, beginning with the lowest leaves, called the "sandlugs," and progressing up the stalk as the leaves ripened. The leaves were taken to a barn where they were separated into two to three dozen groups of three or four, each called a "hand." They were then tied to tobacco sticks with cotton twine and hung in tobacco barns to be cured. In what is known as flue-cured tobacco, the curing process took about a week. The early barns were fired with wood to provide heat to cure the leaves. Later kerosene and gas-fired burners were used.

The curing tobacco had to be carefully monitored. Tobacco farmers had to learn to survive on very little sleep, since they would have to check the barns every few hours, day and night. Too much heat applied too soon would cause an undesirable change in the color of the leaves, lowering their market value. Too little heat would not properly cure them, causing them to mold and rot. If a stick broke and dumped the nearly cured tobacco on a burner, a fire could break out, destroying the barn and its contents and the profit for the season in just a few minutes.

The growers of cigar-wrapper leaf tobacco had some additional duties to perform to produce their crop. In addition to the preparation of the seedbeds and land, the fields where the crop would be grown had to have a framework of poles and laths built over them to support a covering of cheesecloth. This "shade tobacco" was an extremely delicate and valuable crop. The cheesecloth shade was used to keep the leaves from sunburn, and

A true mechanical tobacco harvester. The leaves are removed from the plant by the machine, not by hand.

PHOTO COURTESY OF CLIFF STARLING

Above: A sheet of tobacco weighs approximately two hundred pounds. When tied up in burlap, it is ready to go to market.

Photo courtesy of Starling Farms

As the cured tobacco is removed from the barn, it is packed into a "sheet" of tobacco using this round metal form. The tobacco is then wrapped in a burlap sheet to take to market.

Photo courtesy of Starling Farms

the subdued lighting caused the leaves to become thinner, making a better cigar wrapper. The shade tobacco growers' greatest fear was probably a hailstorm. The hailstones could punch holes in the leaves if they got through the cheesecloth covering, rendering them worthless. Each leaf of shade tobacco was handled with the utmost care and concern. The nearer to perfect each leaf was increased its value. Wrapper leaf tobacco is air-cured, that is, no additional heat is added as in the flue-cured process. When cured, the tobacco went to the markets at Quincy and Havana, to be shipped south to feed the insatiable demands of the cigar rollers in Ybor City, a section of Tampa.

Wrapper leaf could not be adapted to mechanical harvest. Foreign competition, and the widespread use of a homogenized wrapper by the cigar companies doomed the industry. In Florida, it was all but gone by the 1970s.

This trailer load of tobacco is on the way to market.
PHOTO COURTESY OF STARLING FARMS

Farmers watch as buyers from the tobacco companies bid on each individual sheet of tobacco on the warehouse floor. The auction markets are gone now and farmers sell directly to the tobacco companies.

Photo courtesy of Shirley Carte

Flue-cured tobacco has survived. However, as in all things, the tobacco industry has made great changes. Chemical and biological controls now keep many pests and diseases in check. Chemical ripeners are applied that allow the crop to be harvested by machinery. Bulk-curing processes have eliminated the old-style "stick" barns, and the grower markets directly to the tobacco companies at a negotiated price rather than at an auction market.

The machinery and equipment necessary to produce tobacco has all but eliminated the small grower, consolidating the crop into fewer hands, each with larger acreages. While overall production of tobacco is declining, there are more than four thousand acres in production, producing over 11 million pounds, with a cash value around $24 million.

CHAPTER 21

BEEKEEPING

Florida honeybees never get a break. There is always something blooming, always something needing pollination. Florida is one of the four major beekeeping states, and the bees travel all over. Bees are found in citrus groves during the blooming period, hard at work producing world-famous orange blossom honey. Beekeepers rent and place hives near crops requiring pollination. Production of watermelons, cucumbers, and other crops would not be possible without bees to provide essential pollination. The value of bees for pollination exceeds the value of their honey by at least a hundredfold.

Honey consumers all have their favorite kind of honey. Some favor orange blossom, some gallberry, others wildflower, and some find tupelo honey their favorite. Each region of the state has its own distinctive honey.

The citrus honeys can only be produced in the areas that grow citrus. Citrus honey tends to be very mild flavored, very light in color, with an

The beekeeper must carefully smoke and open the hive to prevent the bees from becoming aggressive and excited, which is stressful on the colony.

PHOTO COURTESY OF DALTON TUPPER

aroma that is reminiscent of citrus in bloom. Those used to the mild clover honey produced in other states tend to gravitate to citrus honey when selecting a Florida honey.

Wildflower honey can be produced anywhere. Wildflower is much more variable in flavor, since the bees collect nectar from whatever happens to be blooming. It is always interesting for the wildflower honey producer to see what new flavors the bees have come up with when he extracts a new super of honeycomb.

Some woodlands of North Florida contain an understory plant known as the gallberry bush. The gallberry is not a very distinctive plant, likely to go unnoticed except when it is blooming and the honeybees are collecting nectar. Adherents of gallberry honey claim there is none better tasting, or healthier to eat. However, everyone with a favorite type of honey makes this claim.

The tupelo honey producer has a somewhat different task to get his bees to the blooms. Tupelo only grows along certain rivers in the Panhandle of Florida. To get the bees there, beekeepers have had to put their beehives on barges, and build platforms to keep the hives above water. Tupelo honey has a distinctive flavor, and has the reputation of not crystallizing as other types of honey do.

Beekeepers face a number of hardships to stay in business. There are parasites, pests, and diseases that affect bees and must be guarded against. In those areas where the Florida black bear is found, beekeepers have to take precautions to make their apiaries bear-proof to keep their hives from being raided. The threat of the Africanized honeybee is taken quite seriously in Florida. With so many international ports, both sea and air, special vigilance is necessary to prevent this invader from gaining a foothold in Florida. Port inspectors have found cargo containers with bee colonies in them that have proven to be Africanized bees. The

The beekeeper has removed a frame from the brood chamber. This is where the queen lays her eggs and the young bees are hatched. The beekeeper is checking the brood comb to see to the health of the hive and to determine if it is time to replace the queen.

PHOTO COURTESY OF DALTON TUPPER

Italian-type honeybee that is the standard for beekeepers in this country is quite gentle and easy to handle in comparison to the Africanized type. Many beekeepers currently tend their bees without wearing gloves, and some don't even wear a veil, using only a smoker to calm the bees. This is not possible with the Africanized bee that is ill-tempered, fiercely protective of the hive, and tends to attack in swarms if disturbed. The beekeeper must wear a protective suit, veil and gloves, and a much larger smoker is necessary, and the bees must be smoked more. Life will become much more difficult and dangerous for the beekeeper if the Africanized bee becomes established in Florida.

A veteran beekeeper uses a frame of honeycomb to explain how honey is produced.

PHOTO COURTESY OF HOLT FARMS

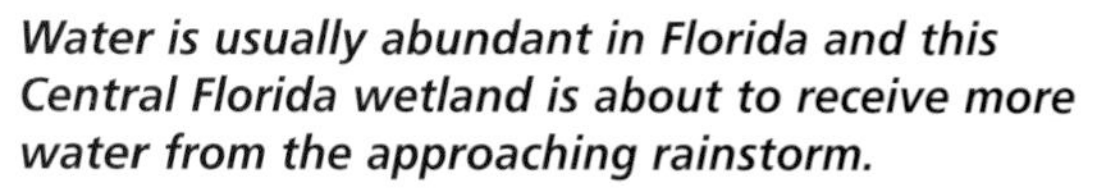

Water is usually abundant in Florida and this Central Florida wetland is about to receive more water from the approaching rainstorm.

PHOTO BY TONY LAYNE COURTESY OF DESERET CATTLE AND CITRUS

CHAPTER 22

NATURAL RESOURCES

Florida is rich in natural resources. While the early explorers may have arrived hoping to find gold, they found instead many natural resources that far exceeded the value of gold. Florida's timber resources attracted the maritime nations to our shores for the lumber and naval stores necessary to build and maintain wooden ships. Later generations discovered vast deposits of phosphate and limestone. Sand, peat, and clay have also been mined from various lands within the state. And as if to fulfill the dreams of those first explorers, some gold has actually been discovered. Small amounts of it have been found in some streams and land. Well drillers in Gainesville found a vein of gold bearing quartz while drilling a new well on the courthouse square in 1897, but analysis found it was of low yield and at an unprofitable depth to mine. Florida's gold has been of a different nature.

Underground water is plentiful. The Florida aquifer comes to the surface in some places, as this spring on the Suwannee River shows.

PHOTO COURTESY OF SUWANNEE RIVER R C & D

The limestone that underlies most of Florida comes almost to the surface in many places, close enough in fact for there to be numerous open pit mines and quarries in areas of North Central Florida. This limestone is quarried and crushed to provide material for roadbeds under paved roads, and to surface unpaved roads. This limestone is also mined and processed for cement, to make fertilizers and soil amendments, and for other industrial uses.

Florida is one of the major producers of phosphate, supplying 25 percent of the phosphate needs of the world. Only three other states mine phosphate. Florida mines produce 75 percent of the phosphate needed in the United States. The first discoveries of phosphate in the United States were in Florida, in 1881, along the Peace River where mining initially began. Some mining has also been conducted in the northern part of the state. There was once a considerable industry mining the "hard rock" phosphate around the towns of Newberry and Archer in Alachua County, and around Dunnellon in Marion County. Much richer deposits of "soft rock" phosphate were discovered in Polk and Hillsborough Counties, and the bulk of the mining of this essential mineral is conducted there. Phosphate is used in the production of commercial fertilizer. It provides phosphorus, the "P" in NPK, the three major plant

Diverse lands keep us from "putting all our eggs in one basket." Florida soils vary greatly from one end of the state to the other. This is typical rolling sandy loam soil found in the Panhandle and North Central Florida.

PHOTO COURTESY OF FRANKIE ALLEN

Florida's forestlands are an abundant resource. Properly managed they will continue to provide lumber, paper, and a host of other products.

PHOTO COURTESY OF R. J. NATHE AND SONS

nutrients: nitrogen, phosphorus, and potassium. In the United States, fertilizer is formulated around these three elements. The label on a bag of fertilizer will list the product as 3-9-18, 5-10-15, 8-8-8, and so on. These numbers indicate the percentage of the elements nitrogen, phosphorus, and potassium in the formulation. It's almost a sure bet that the phosphorus indicated by the middle number was mined in Florida.

Florida's greatest natural resources are its land and water. Floridians are proactive in protecting these irreplaceable elements.

Programs are in place to minimize the loss of topsoil to erosion and oxidation. Many Florida farmers are voluntarily utilizing best management practices in fertilization, chemical application, irrigation, and animal waste disposal. These practices protect the land and the water, and are economically advantageous. Florida farmers believe that voluntary self-regulation is preferable to mandated regulation.

Florida has some of the best water resources anywhere in the world. There are countless lakes, streams, and rivers throughout the state. In many places the water table is at or near the surface. The basin of the Suwannee and Santa Fe Rivers contains the greatest concentration of freshwater springs on the planet. These cold, crystal-clear springs are popular places for recreation. Swimming, snorkeling, and cave diving are all activities that residents and tourists alike participate in. Some springs are so large

Soils farther south can be considerably different. This scene of rice harvest in the Everglades shows that Florida has soils that support a variety of crops.
PHOTO COURTESY OF HOLT FARMS

that they create crystal-clear rivers as their outlets. Some examples of these are the Ichetucknee, the Rainbow, and the Silver Rivers. Probably the best known of these freshwater springs is the famous Silver Springs, known nationwide as a tourist destination. Many springs are located in less developed areas, and have a large local following. Many Florida families have been visiting the same springs for generations. Both state agencies and local governments have acquired a number of these springs and the land surrounding them, both to protect them, and to provide places for outdoor recreation.

With these many water resources, and with the Florida aquifer so close to the surface, it is of prime importance to maintain the highest level of water purity. As municipal, residential, and agricultural consumers all draw water from the underground aquifer, any contamination would have a widespread and devastating effect.

The caracara is also called the Mexican buzzard.
PHOTO COURTESY OF LIGHTSEY CATTLE COMPANY

CHAPTER 23

WILDLIFE

Florida wildlife is one of its most valued natural resources. Both game and nongame species abound. Deer herds and wild turkey flocks are at an all-time high. Wild hogs roam the woodlands. Florida has the largest population of bald eagles outside of Alaska. Because it is so far south, migratory birds come by the millions to spend the winter in Florida. Some of the wildlife is unique and foreign to the rest of the country. The fearsome-looking alligator is found in large numbers, and there is a species of crocodile found in the Everglades. The black bear has a home in some of the northern counties. The elusive Florida panther is found in several locations in the state. The bobcat is found throughout the state. There are some animals that are found nowhere else, such as the diminutive Key deer.

Florida is assumed to be a land of snakes and lizards. To a certain extent that is true. There are more species of snakes and lizards in Florida than

The Florida alligator is more abundant now than ever. They can become a nuisance when they blunder into swimming pools in housing developments, and become dangerous when fed. Left alone in their natural habitat, they are a valued part of the ecosystem.

PHOTO COURTESY OF LIGHTSEY CATTLE COMPANY

in any other state in the continental United States. All but a very few snakes are nonvenomous, and there are no venomous lizards in Florida. Both the venomous and nonvenomous species tend to be nonaggressive. Some like the indigo snake make excellent pets, and thus have become endangered in the wild. The indigo snake and others have found some of their last refuges in the woodlands and fields of Florida.

Reptiles like the gopher tortoise have come under increased pressure because of habitat loss. The continual development of the

Gobblers strut to impress the hens in this flock of turkeys.

PHOTO COURTESY OF LIGHTSEY CATTLE COMPANY

The American bison, or buffalo, once roamed Florida. Now only a few, like this cow and calf, remain on park and private lands in the state.

PHOTO COURTESY OF LIGHTSEY CATTLE COMPANY

farm and woodlands of the state, and the subsequent loss of habitat and isolation of populations, have also made Florida a state with a number of endangered, protected, and species of special concern.

Florida's tropical, semitropical, and temperate climate zones are conducive to a number of nonnative species that have become established here. The white cattle egrets that are seen with livestock in every part of the state are native to West Africa, and their presence here is an unsolved mystery. One theory is that a flock of them got trapped in the eye of a hurricane and flew with the storm all the way across the Atlantic, settling here after the storm made landfall. All that is known for sure is that they

Fallow deer from England have been introduced on some private game preserves in Florida.

Photo courtesy of
Lightsey Cattle Company

While not wildlife, a hog-and-cow dog like Gus, a blackmouth cur, is indispensable for catching the wild hogs that are in abundance on some lands in Florida.

Photo courtesy of
Lightsey Cattle Company

An eastern diamondback rattlesnake is venomous, but fortunately not very aggressive and will run from humans if given the chance.

Photo courtesy of Lightsey Cattle Company

The white head that is distinctive in the mature bald eagle has not yet developed on this immature bird. This one is feeding on the remains of some other species of Florida wildlife.

Photo courtesy of Lightsey Cattle Company

appeared here in the 1920s and have been here since. The coyote is another animal that has made a home here, and in some areas causes serious depredation to livestock. Numerous other animals, fish, and shellfish inhabit the state, some introduced deliberately, many more inadvertently.

Plants and trees have found their way here, some on their own, some with help. The Australian melaluca tree was introduced to help with drainage projects, and has since become a pest. Kudzu was brought from China to be used as cattle forage, and now covers and suffocates trees and other plants. Water lilies were introduced into lakes and rivers and now choke some bodies of water. Hydrilla was a harmless aquarium plant until it was introduced into the wild. It now makes boating and fishing on some lakes difficult, if not impossible. There are several undesirable grasses that threaten to supplant desirable pasture and lawn grasses. Numerous annual and perennial weeds have made their way to Florida and create difficulties for the farmer, rancher, and homeowner.

A mature bald eagle guards a nest on Brahma Island, Lake Kissimmee.

PHOTO COURTESY OF LIGHTSEY CATTLE COMPANY

CHAPTER 24

PEOPLE IN AGRICULTURE

PAST, PRESENT, AND FUTURE

The strength of Florida agriculture lies in its people. Florida farmers and ranchers have overcome many obstacles, both natural and manmade, to thrive and prosper. People from every nation on earth have come to the state to follow their dreams and sink roots in the fertile soil of Florida. More than one "rags to riches" story can be told about Florida. Empires of cattle, citrus, and vegetables have been built by people who came here with little more than the clothes on their backs, but with a will and drive to succeed. This attitude is still to be found. The wilderness and all of its dangers are gone. The raw land that the early settlers and Native Americans faced has been tamed. The open rangelands that once stretched the length of the state have been fenced.

The frontier of knowledge will always remain. The challenges of production have not diminished in spite of technology. The farmer and rancher of

Retired vocational agriculture teachers dine at their annual reunion at O'leno State Park, home of the State Forestry Camp.

Photo courtesy of Joe Kirkland

tomorrow must be better educated, more willing to utilize cutting-edge technology, and willing to work just as hard as his forebears did.

The future of Florida agriculture lies in the next generation. Those who will succeed the current generation are already working on their career choices. The future will see what farmers, ranchers, county agents, researchers, processors, marketers, and teachers will step forward to continue the work that has gone on for centuries.

Don McCormick, a retired teacher, supervisor, and vocational director, emcees the program at the annual meeting of the Seasoned Owls, an organization of retired teachers.

Photo courtesy of Joe Kirkland

More than one generation of Florida cowboys is in this group.

PHOTO COURTESY OF LIGHTSEY CATTLE COMPANY

The home of the Florida FFA Foundation is at the Florida FFA Leadership Training Center.

PHOTO COURTESY OF GARY BARTLEY

This is the marina at the Florida FFA Leadership Training Center.

PHOTO COURTESY OF GARY BARTLEY

Young people involved with the FFA and 4-H have traditionally come from an agricultural background. That, like all things, has changed. More and more members of these organizations are coming from urban and suburban settings, with limited or nonexistent exposure to agriculture in the home setting. The core values of leadership, hard work, and responsibility that these programs instill are just as valid today as they were in the past. Through the hard work of FFA advisors and 4-H leaders, members learn that nothing is worth having that is not worked for, that the greatest satisfaction comes from accomplishing a task that seems daunting at first, that teamwork is the best way to accomplish a difficult task, that confidence is gained through experience, and that the ability to grasp an opportunity is the way to get ahead.

FFA and 4-H members have the benefit of the experience of previous generations to guide them. People who wore the blue jacket or the green jacket never forget their experiences, including the opportunities they enjoyed.

Former members have banded together to form support organizations to help provide new opportunities for today's young people. Florida FFA members are supported by a network of alumni chapters, each with the goal of finding ways to assist the local FFA chapters in providing additional educational experiences, through travel, competition, leadership activities, and scholarships.

The Florida FFA Foundation has existed for over a quarter century with the goal of providing funding to support leadership and educational activities for FFA members. An impressive facility, the FFA Leadership Training Center has been built on the shore of Lake Pierce, near Haines City. This facility is utilized year-round by FFA members, as well as by leaders of business and industry.

Florida FFA officers visit a Seminole Feed plant during their annual goodwill tour.
PHOTO COURTESY OF ERIN BEST

Students participate in a land judging and homesite evaluation contest, a highly technical and difficult contest.
PHOTO BY THE AUTHOR

One of the guest lodges at the Leadership Training Center.

PHOTO COURTESY OF
GARY BARTLEY

4-H is supported by the Florida Cooperative Extension Service through its network of extension offices in each county. 4-H leaders coordinate their activities and obtain resources through these offices.

A number of counties have groups with names like "Friends of 4-H and FFA" made up of community leaders and other citizens who are actively involved throughout the year promoting and supporting the activities of these two outstanding organizations.

One of the next generation of Florida cowboys gets used to the saddle.

PHOTO COURTESY OF
LIGHTSEY CATTLE COMPANY

The future begins right now. The past is merely a memory. The present slips away so fast that it is gone before we realize it. We can choose, just like those who came before us, to build a better tomorrow. The rich legacy that our forefathers left is not to be discarded in an idle or thought-less moment. The trials and tribulations of previous generations can never be forgotten lest we disgrace their memory and sacrifice. The responsibility of leaving the world a better place for those yet unborn is ours alone. No government program, no technological breakthrough, will do it for us. Nothing has ever been accomplished without a lot of hard work. The words from the opening ceremony of the FFA say it best, "Without labor, neither knowledge nor wisdom can accomplish much."

A young competitor communes with her show hog.

PHOTO COURTESY OF ERIN BEST

It is never too young to start, but a little help from mom sure is nice. This young competitor won't need help for long. Independence and self-reliance are hallmarks of 4-H and FFA competition.

PHOTO COURTESY OF LARRY AND DEBBIE SWINDLE

Bibliography

Alachua County Clerk of Court. *Brands.* Gainesville: Alachua County, 1919–1956.

Alachua County Clerk of Court. *Inventories and Appraisements 4.* Gainesville: Alachua County, 1858–1880.

Florida Department of Agriculture and Consumer Services, Division of Marketing and Development: *Florida Agricultural Facts 2002.*

Gannon, Michael. *Florida: A Short History.* Gainesville: University Press of Florida, 1993.

Gerry, Eloise. *Improvement in the Production of Oleoresin through Lower Chipping.* Washington, D.C.: United States Department of Agriculture, 1931.

Getzen, Sam P. *Crackers: Florida's Heritage Horses.* Newberry: The Florida Cracker Horse Association, 2002.

Harris, Wayne. *Florida's Agricultural Heritage.* Tampa: Florida Agricultural Hall of Fame Inc., 2002.

Mattoon, Wilbur R. *Longleaf Pine.* Washington, D.C.: United States Department of Agriculture, 1925.

National FFA Organization. *Official Manual.* Indianapolis: National FFA Organization, 2002.

Poertner, Bo. *Old Town by the Sea: A Pictorial History of New Smyrna Beach.* Virginia Beach, Va.: The Donning Company Publishers, 2002.

Sampson, R. Neil. *For Love of the Land.* League City, Texas: National Association of Conservation Districts, 1985.

Index

About the Author

A native of Alachua County, Archie Matthews has been a longtime supporter of the FFA. He was a member of the Santa Fe FFA chapter, and continues to serve the chapter as an Alumni member. He has served on the Board of Directors of the Florida FFA Alumni Association for most of the history of the organization, and has served five terms as president. In 2001 he was elected to the office of Director Emeritus by the Alumni Board of Directors. He and his wife Emelie support youth activities through their church, the FFA, and student organizations at the University of Florida. He is employed as Archives Manager for the Alachua County Clerk of Court. Along with son David, he manages the family farm, a timber and cattle operation in western Alachua County.